Resisting Russia

Insights into Ukraine's Civilian-Based Actions During the First Four Months of the War in 2022

MARTA KEPE, ALYSSA DEMUS

Prepared for the Office of the Secretary of Defense
Approved for public release; distribution unlimited

For more information on this publication, visit **www.rand.org/t/RRA2034-1**.

About RAND

The RAND Corporation is a research organization that develops solutions to public policy challenges to help make communities throughout the world safer and more secure, healthier and more prosperous. RAND is nonprofit, nonpartisan, and committed to the public interest. To learn more about RAND, visit www.rand.org.

Research Integrity

Our mission to help improve policy and decisionmaking through research and analysis is enabled through our core values of quality and objectivity and our unwavering commitment to the highest level of integrity and ethical behavior. To help ensure our research and analysis are rigorous, objective, and nonpartisan, we subject our research publications to a robust and exacting quality-assurance process; avoid both the appearance and reality of financial and other conflicts of interest through staff training, project screening, and a policy of mandatory disclosure; and pursue transparency in our research engagements through our commitment to the open publication of our research findings and recommendations, disclosure of the source of funding of published research, and policies to ensure intellectual independence. For more information, visit www.rand.org/about/principles.

RAND's publications do not necessarily reflect the opinions of its research clients and sponsors.

Published by the RAND Corporation, Santa Monica, Calif.

Library of Congress Control Number: 2023912657

ISBN: 978-1-9774-1138-9

Cover: Reuters photo.

About This Report

The ongoing war in Ukraine has highlighted the many ways in which civilians can support a nation's effort to defend itself against an external occupying power. Although civilian-based resistance began during the first hours of the Russian invasion of Ukraine on February 24, 2022, most analysis to date has focused on Russian-Ukrainian armed military confrontations. Ukrainian *civilian-based resistance efforts* merit attention because of their potential value in helping Ukraine's strategic aims to ensure victory by regaining territorial integrity and maintaining political sovereignty. The war in Ukraine also offers some insight into potential future trends in civilian actions in a conventional war for territorial integrity. In this report, we examine Ukrainian approaches to civilian-based resistance during the first four months of the Russia-Ukraine War of 2022 using an analytical framework previously developed by RAND Corporation researchers to analyze civilian-based efforts to support national resistance against a foreign occupying power. Because the war is ongoing and information about many pertinent civilian-based activities in support of the military might not yet be publicly available, we offer a broad characterization of civilian resistance in Ukraine. Therefore, this report is not intended to be an exhaustive analysis of civilian-based resistance in Ukraine. Using this analysis, we identify resistance activities that could be indicative of future trends in civilian-based resistance against external aggressors and potential areas of future research and analysis.

The research reported here was completed in November 2022 and underwent security review with the sponsor and the Defense Office of Prepublication and Security Review before public release.

RAND National Security Research Division

This research was conducted within the International Security and Defense Policy Program of the RAND National Security Research Division (NSRD), which operates the RAND National Defense Research Institute (NDRI), a federally funded research and development center (FFRDC) sponsored by the Office of the Secretary of Defense, the Joint Staff, the Unified Combatant Commands, the Navy, the Marine Corps, the defense agencies, and the defense intelligence enterprise. This research was made possible by NDRI exploratory research funding that was provided through the FFRDC contract and approved by NDRI's primary sponsor.

For more information on the RAND International Security and Defense Policy Program, see www.rand.org/nsrd/isdp or contact the director (contact information is provided on the webpage).

Acknowledgments

We would like to thank Mark Cozad, Jim Mitre, Heather Williams, and Anika Binnendijk for their support and guidance. We are grateful to our peer reviewers Stephen Flanagan and Pauline Moore.

Summary

In this report, we examine the contributions that Ukrainian civilians have made to support Ukraine's defense against the Russian invasion since February 2022, with a specific focus on the first four months of the war. To better understand the civilian-based resistance efforts in 2022, we also offer a short overview of the most-recent history of civilian-based resistance and opposition movements that have paved the way for stronger social mobilization and activism across all segments of society.

Civilian-based activities are an integral part of Ukraine's territorial defense efforts. To consider how Ukrainian civil society has supported its country's national defense, we adopted the five proximate objectives for civilian contributions to resistance set forth by an analytical framework previously developed by RAND Corporation researchers. These objectives were specifically created to analyze civilian-based resilience preparations and resistance against a potential foreign aggressor in the Baltic states and were published in the 2021 report *Civilian-Based Resistance in the Baltic States: Historic Precedents and Current Capabilities*.[1] These objectives are

1. imposing direct or indirect costs on an occupying force
2. securing external support
3. denying the occupier's political and economic consolidation
4. reducing the occupier's capacity for repression
5. maintaining and expanding popular support.

Our report is affected by certain limitations, including the inherent challenge of researching an ongoing war and the limited availability or precision of openly published information. As a result, our report offers some preliminary insights into civilian-based activities supporting Ukraine's national defense. We also observe several trends that might be relevant for future civilian-based resistance campaigns against a foreign occupying power, including cyber wars, civilian skills in support of national resistance, the

[1] Anika Binnendijk and Marta Kepe, *Civilian-Based Resistance in the Baltic States: Historic Precedents and Current Capabilities*, RAND Corporation, RR-A198-3, 2021.

adversary's ability to consolidate its economic power, the role of large businesses in wars, and the significance of prior social mobilization experiences.

Key Findings

Our key findings include the following:

- Civilian-based activities in Ukraine in February–June 2022 were varied and included numerous actors, ranging from civilian government actors and political leadership to independent enterprises and self-organized citizen groups.
- Many civilian-based activities were spontaneous, need-based, and relied on existing informal networks at the community level, while higher-level coordination was more present in civil protection, humanitarian aid, hacker activism, and communication activities.
- The Russia-Ukraine War of 2022 provides a glimpse into what future civilian contributions to wars could look like. Interstate conflict could involve large-scale cyber operations waged by a loose confederation of amateurs and professionals from the occupied country and abroad.
- Civilians with specialist and generalist skills might offer meaningful contributions by imposing direct military costs on occupying powers and supporting the defending armed forces or civil society. When considering potential military conflict, it is necessary to prepare guidelines for such eventualities to avoid instances in which a sudden surge of civilian volunteers creates chaos, causes friendly fire, or interferes with the strategic objectives of the country.
- The ability to deny an occupier's economic consolidation of the occupied country will increasingly mean protecting not only key economic centers of gravity during the war but also peacetime resilience preparations, including the diversification of energy resources and other product supply chains and—if the country is a significant exporter—export lines, especially in the unoccupied territories.
- Ukraine's history of social and political movements helped build a stronger and more active society; this history shows the value of understanding civil society movements in different countries. In-depth study

of the history of national civil resistance movements within countries located in potential conflict areas might help the United States, its allies, and partners develop a better understanding of how these countries conceptualize civil resistance, trends in civil-resistance methods, and their application to current or future conflicts.

- In future wars, large companies could increasingly shape the operational aspects of a conflict and the geopolitical and strategic outcomes of a conflict.

Contents

Tables

CHAPTER 1

Introduction

Since February 2022, Ukraine has been engaged in a battle to defend its sovereignty and territorial integrity in the face of a major Russian invasion. Although global audiences have been focused on the battlefield, civilian-based resistance activities have emerged as another significant yet sometimes less visible layer of Ukraine's territorial defense and societal resilience against Russian attempts at conquest. Civilian-based activities have garnered little attention from scholars, practitioners, and the media.[1] Nevertheless, past research on civilian-based resistance broadly indicates—and our initial exploration of this conflict specifically supports—that civilian resistance against external occupation or forces is a critical component of a state's defense against a foreign occupying power, including in Ukraine today.

Defining Civilian-Based Resistance

By *civilian-based resistance*, we mean a "nation's organized whole-of-society effort, encompassing the full range of activities from nonviolent to violent, led by a legally established government . . . to reestablish independence and autonomy within its sovereign territory that has been wholly or partially

[1] For a rare example, see Felip Daza Sierra, "Ukrainian Nonviolent Civil Resistance in the Face of War: Analysis of Trends, Impacts and Challenges of Nonviolent Action in Ukraine Between February and June 2022," *International Catalan Institute for Peace and International Institute for Nonviolent Action*, 2022.

occupied by a foreign power."[2] Although embroiled in a war with Russia as we write, Ukraine remains an independent and sovereign country, with only some of its territory occupied by Russian forces. Therefore, while we situate our discussion within the analytical framework that is designed for the analysis of resistance against foreign occupation, we acknowledge that the civilian-based activities in Ukraine are taking place as an integral part of Ukraine's territorial defense efforts and are designed specifically to counter Russian forces in the occupied territory.

Before delving into the specifics of what is (and is not) included in this framework, it is important to acknowledge that existential conflicts (such as the one this short report examines) are complex and messy; often, it may be difficult to clearly classify people and organizations strictly as "civilian" or "combatant." Thus, although we aim to explicitly differentiate among those individuals, organizations, and activities that our analytical approach includes, some cases do not fall neatly into one category.

This report examines the contributions of civilian populations to Ukrainian resistance during the first four months of the Russia-Ukraine War of 2022. We adopt the Binnendijk and Kepe (2021) analytical framework that was initially developed to analyze civilian-based resistance and resilience preparations against a potential foreign aggressor in the Baltic states.[3] Specifically, this construct includes the contributions of civilian populations and entities to national resistance, meaning activities undertaken by (1) individual civilians (and groups of civilians) unaffiliated with any formal government entities, (2) for-profit and nonprofit civilian organizations, and (3) civilian (nonmilitary) elements of the Ukrainian government. Regarding the activities covered in our construct, they span (1) direct and indirect unarmed and armed action against the adversary at the hands of civilian entities, (2) various means of soliciting support from external sources, and (3) measures aimed at delivering aid and protection to local civilians (see Table 1.1).[4]

---

[2] Otto Fiala, *Resistance Operating Concept*, Swedish Defense University and Special Operations Command Europe, 2019, p. 15.

[3] Anna Binnendijk and Marta Kepe, *Civilian-Based Resistance in the Baltic States: Historic Precedents and Current Capabilities*, RAND Corporation, RR-A198-3, 2021.

[4] Binnendijk and Kepe, 2021.

Our approach to civilian-based resistance includes *all* efforts carried out by civilians—meaning noncombatants—but not those of civilians who voluntarily join the Ukrainian Armed Forces or Territorial Defense Forces. Therefore, this report considers efforts carried out by civilians that fall wholly outside formal government structures and civilians involved in nonmilitary efforts at the direction of government authorities. We also include efforts that are designed and/or conducted by civilian entities who are directed by—or at least coordinated with—official civilian government authorities, the Ukrainian Armed Forces, or the Security Service of Ukraine (SBU). One notable example is civilian-based information-gathering on enemy positions.[5] The civil–military links developed by civilians who have joined the Ukrainian Armed Forces enable military and security forces to solicit reconnaissance information from local civilians. Some illustrative examples include reporting by women who distribute pensions by visiting the elderly and by beekeepers about the location of their beehives to the Ukrainian authorities.[6]

Lastly, we also subsume activities planned and executed without the involvement of any authorities at the national or local levels into our conception of civilian-based resistance, including protests, noncooperation with Russian forces, and informal volunteer networks of aid delivery to civilians.[7]

Drawing on the extensive literature on civil resistance, the Binnendijk and Kepe framework identifies five proximate objectives "that conceptually link the tactics and actions of civilian-based campaigns to the desired resistance outcome: independence from occupation."[8] These proximate objectives are outlined in Table 1.1.

Notably, neither the framework nor this report includes activities conducted by civilians who engage in military actions by joining the Territorial Defense Forces or by volunteering for mobilization. That said, a related component, and one that is a key civilian contribution to Ukraine's mili-

[5] Norma Costello and Vera Mironova, "Ukraine Has a Secret Resistance Operating Behind Russian Lines," *Foreign Policy*, November 21, 2022.

[6] Costello and Mironova, 2022.

[7] Daza Sierra, 2022.

[8] Binnendijk and Kepe, 2021, p. xi.

TABLE 1.1
Proximate Objectives

Proximate Objectives	Potential Contribution to Success	Relevant Tactics and Actions
Imposing direct or indirect costs on an occupying force	• Make it untenable or undesirable for occupying power to remain	• Increase personnel and material requirements through sabotage, armed force, or noncooperation • Elicit international sanctions • Target domestic political audiences in aggressor state
Securing external support	• Increase international pressure to withdraw • Enhance resources for resistance • Apply direct costs through military intervention or sanctions	• Engage in targeted outreach • Leverage international institutions and relationships • Apply communications strategy
Denying an occupier's political and economic consolidations	• Preserve legitimacy and function of occupied government • Deny adversary political or economic support	• Establish government in exile • Maintain control of resistance to ensure rule of law and legitimacy • Prevent co-option of economic centers of gravity
Reducing an occupier's capacity for repression	• Establish backfire to erode occupier legitimacy[a] • Preserve popular will to resist	• Relocate vulnerable populations • Publicize acts of repression • Clearly delineate noncombatants
Maintaining and expanding popular support	• Broaden base and resilience of resistance • Preserve popular will to resist	• Create low-risk opportunities for participation • Develop and maintain coherent message • Establish effective communication systems

SOURCE: Reproduced from Binnendijk and Kepe, 2021, p. 12.

[a] *Backfire* refers to a dynamic that can occur when acts of violence by an occupying force, if exposed, elicit outrage within critical audiences and prompt them either to impose costs on the occupier or to provide the resistance with other forms of support.

tary defense efforts, is the civilian government's establishment of formalized paths that allow civilians to join the armed fight. The Ukrainian government has worked to institute such mechanisms by creating the Territorial Defense Forces (*Viiska Teytorialnoi Oborony*), which was based on the previous looser structure of the Ukrainian Territorial Defense Battalions and volunteer militias that were instrumental in the Russia-Ukraine War in Eastern Ukraine in 2014–2022. Reports suggest that other town- or community-based self-organized defense groups have also emerged since the February 2022 Russian invasion. One such example is the Bucha Community Defense Force, which is a locally focused unit financed by the local community instead of the central government.[9] The civil–military links developed when civilians join the Ukrainian Armed Forces, or when veterans coordinate with their former counterparts, contribute to the ability of the military and security forces to solicit reconnaissance information from local civilians.[10]

We also acknowledge that direct action has attracted foreign volunteers who have joined the hastily created International Legion of Territorial Defense of Ukraine—a tool that attracts foreign veterans with recent combat experience and shows that, in the words of Dmytro Kuleba, "The whole world today is on Ukraine's side, not only in words but in deeds."[11] The activities of these foreign volunteers also fall beyond the scope of this short report.

We base our analysis in this report on a review of openly available materials: official documents, analytical reports, media reports, and information available on social media in English, Ukrainian, Swedish, and Russian. We limited the scope of our research to the first four months of the Russia-Ukraine War, February–June 2022, and we finalized our research in November 2022. Therefore, this report reflects the information available at that time.

---

[9] Liam Collins and John Spenser, "In Ukraine, Two US Army Veterans See the New Face of War," *Soldier of Fortune Magazine*, August 22, 2022.

[10] Costello and Mironova, 2022.

[11] Collins and Spenser, 2022; and Robin Wright, "Will Mercenaries and Foreign Fighters Change the Course of Ukraine's War?" *New Yorker*, April 5, 2022.

Report Structure

To contextualize and situate Ukraine's civilian-based resistance efforts in the country's broader history, we open with a brief overview of the history of resistance in Ukraine. We then examine the Ukrainian civilian-based resistance activities in 2022 through the lens of the Binnendijk and Kepe framework. This report is not intended to serve as a comprehensive analysis of the ongoing resistance efforts in and around Ukraine. Rather, we offer examples of civilian-based activities that might support each of the five proximate objectives and seek to enhance our understanding of the strength of Ukrainian civilian-based resistance activities and their potential implications for war in the 21st century.

Caveats and Limitations

This publication has certain limitations. First, it is inherently challenging to carry out in-depth research on an ongoing war because publicly available information on the multiple facets of civilian activities may be limited or suppressed. Second, this report is based on a review of publicly available materials. We reviewed numerous publicly available reports, articles, and social media posts, in addition to official Ukrainian government sources. Because of the changing situation on the ground, constantly developing state of information, and potentially imprecise data (for example, on casualty levels and other conflict dynamics), any specific metrics used in this report serve purely illustrative purposes; we aim to develop a general overview of the civilian-based resistance trends in Ukraine. As a result, we also adopt a careful approach regarding any potential conclusions. Because of resource and time limitations, we could not carry out research interviews or travel to Ukraine for more extensive research.

CHAPTER 2

Civilian-Based Resistance in Ukraine's History

Ukrainian resistance efforts throughout the 20th century and Ukraine's more immediate past are particularly significant to civilian-based movements in Ukraine in 2022. The civilian resistance efforts we watched unfold in Ukraine since February 24, 2022, have many precedents throughout Ukraine's long-standing—though at points checkered—tradition of defying external occupiers. Throughout its history, Ukraine has witnessed the emergence of various opposition movements, which mobilized to reject what was then perceived as occupation, even during periods when the Ukrainian people were not organized into a self-governing state.[1] These antecedent efforts—and particularly those of the past two decades—have set precedents for a stronger social mobilization in 2022. Therefore, any attempt to understand the civilian-based aspects of Ukraine's struggle against Russian occupation as part of the ongoing conflict requires at least some exploration of Ukraine's past. This is especially true because the more controversial and more traumatic elements of Ukraine's past have become part of the Kremlin's narrative to justify the war. This section will briefly discuss pertinent examples from the most-recent history of civilian-based resistance in Ukraine, focusing on those chapters in its history during which Ukrainians

[1] In fact, Ukraine has only experienced two, albeit relatively brief, periods of independence as a sovereign state: between 1917 and 1921 and, most recently, since 1991 following the dismantling of the Soviet Union.

mounted opposition campaigns against Russian (and Soviet) occupation or political influence.[2]

In the first two decades of the 20th century, Europe endured several seismic geopolitical upheavals, including the outbreak and conclusion of World War I and the ejection of the Russian monarchy in 1917. Each had significant consequences for the Ukrainian people, including tumult, violence, and shifting boundaries but also the opportunity for self-governance. Several grassroots movements vying for independence erupted among different groups of Ukrainian peoples at the twilight of World War I, the interwar period, and after World War II. In western Ukraine, Galician Ukrainians who aimed to establish an independent Western Ukrainian Republic revolted against Polish occupation in 1918, but their efforts were short-lived.[3] A decade later, however, these same exiled individuals formed the Organization of Ukrainian Nationalists (*Orhanizatsiia Ukrains'kykh Natsionalistiv*), or OUN. Although the OUN was organized around the objective of establishing an independent Ukrainian state, some of its aims and means were controversial. Notably, because the OUN's vision of a Ukrainian state was predicated on the liberation of lands from Polish, Soviet, Czechoslovak, and Romanian hands, the group aligned itself with what it perceived as the only other viable ally, Nazi Germany.[4]

---

[2] There are several important factors complicating any retelling of Ukrainian history. First, as is the case with many peoples in the throes of nation-building, the forebearers of present-day Ukrainians cannot be characterized as a singular group for much of Ukraine's history but rather an evolving mosaic of subgroups that were sometimes fractured along east–west lines. Second, aspects of Ukraine's history have been embellished or cloaked by those documenting and retelling it, deliberately and inadvertently. While acknowledging their contested nature, we discuss these history episodes because of their impact on today's resistance movement.

[3] Timothy Snyder, *The Reconstruction of Nations: Poland, Ukraine, Lithuania, Belarus, 1569–1999*, Yale University Press, 2003.

[4] Notably, the group carried out assassinations of Polish and Russian officials in the name of bolstering "the strength, riches, and size of the Ukrainian State even by means of enslaving foreigners" and expelling all but ethnic Ukrainians from what the group viewed as Ukrainian territory. Although the OUN and its collaboration with the Nazis holds a prominent place in the pages of history books, it is important to note that the OUN only reflected a small minority of Ukrainian views at the time. See Snyder, 2003, pp. 143, 152.

In the Soviet-held parts of what is now central and eastern Ukraine (approximately), the early 1930s witnessed the extermination of millions of Ukrainians in the *Holodomor* genocide (killing by starvation) at the hands of the Soviet authorities and a culling of Ukrainian intellectuals as part of Joseph Stalin's purges in the late 1930s.[5] By the time the dust had settled from World War II, the Soviets had swallowed several western Ukrainian territories, including the western reaches of Ukraine today, and refocused their attention on the Ukrainian Anti-Soviet resistance movements.[6] These movements were no match for the much larger and better equipped Soviet forces, whose efforts to extinguish the Ukrainian resistance movement was successful in hollowing out the OUN by the early 1950s and effectively terminating the group by the end of the decade.

Moscow has weaponized certain elements of this period of Ukraine's resistance history. Russian messaging on its war in Ukraine has been largely predicated on the spurious assertion that the residents of the Donbas region in Ukraine are "liv[ing] under the yoke of the [Ukrainian] neo-Nazi regime," in the words of Vladimir Putin.[7] Therefore, Moscow has leveraged the checkered portions of Ukraine's history—such as the OUN's ties to the Nazi regime—to justify its 2022 invasion of Ukraine, claiming that the authorities in Kyiv are fascists and attempting to besmirch Ukraine's military response and the country's civilian-based actions.[8]

Ukraine's past struggles for independence have shaped modern-day Ukrainian ethos, cultural narratives, and national symbols. In fact, some decades- and even centuries-old instruments of marginalization continue to reverberate in the Ukrainian resistance today. Take the issue of language, for instance. In response to the emergence of a nationalist Ukrainophile movement, which gained ground in the early to mid-1800s, Russian tsarist authorities banned the publication of texts in the Ukrainian language and

---

5 Timothy Snyder, *Bloodlands: Europe Between Hitler and Stalin*, Basic Books, 2010; and Anne Applebaum, *Red Famine: Stalin's War on Ukraine*, Doubleday, 2017.

6 Snyder, 2003, p. 203.

7 "Read Putin's National Address on a Partial Military Mobilization," *Washington Post*, September 21, 2022.

8 "Read Putin's National Address on a Partial Military Mobilization," September 21, 2022.

outlawed the use of the language in schools and other public fora.[9] Now the Ukrainian language is one of the symbols of national resistance against Russian invasion.

Preface to the 2022 Resistance

The twilight years of the Soviet period from 1989 to 1991 witnessed the emergence and maturation of new Ukrainian civic organizations. Many were organized under a common parent organization, the People's Movement of Ukraine, or *Rukh*, which was responsible for convening large-scale demonstrations. These demonstrations brought tens, sometimes hundreds, of thousands of Ukrainians to the streets to protest the Ukrainian Soviet authorities and support noncommunist opposition figures.[10] The events surrounding this period are known as the Granite Revolution, named after the granite pavement stones on the Independence Square (then the October Revolution Square) on which a group of students pitched their tents to protest Communist Party policies.[11] *Rukh's* tenure was short-lived, as its organizational structure began to disintegrate following the collapse of the Soviet Union. Even so, its legacy lives on, as it set the precedent for mass demonstrations in the context of the Orange Revolution in 2004 and the Revolution of Dignity (commonly known as the Euromaidan, or Maidan Uprising) in late 2013–early 2014.

The civil movements in the newly independent Ukraine should be considered in the context of its political environment. The period immediately following the Soviet Union's collapse was profoundly chaotic and tumultuous for Ukraine. Ukraine's domestic political developments and its relationship with Russia were deeply entwined. Although Ukraine formally declared

---

9 George O. Liber, *Total Wars and the Making of Modern Ukraine, 1914–1954*, University of Toronto Press, 2016, p. 24.

10 Adrian Karatnycky, "The Fall and Rise of Ukraine's Political Opposition: From Kuchmagate to the Orange Revolution," in Ander Åslund and Michael McFaul, eds., *Revolution in Orange: The Origins of Ukraine's Democratic Breakthrough*, Carnegie Endowment for International Peace, 2006, pp. 29–30.

11 Coilin O'Connor and Halyna Tereshchuk, "The Revolution on Granite: Ukraine's 'First Maidan,'" *Radio Free Europe/Radio Liberty*, October 15, 2020.

its sovereignty, followed by its independence in 1991, in practice, the longstanding overt and surreptitious ties binding Kyiv to Moscow were not so easily severed. Chief among the formal linkages was Ukraine's dependence on Russian energy, which persisted for decades following the collapse of the Soviet Union.[12] Behind the scenes, Moscow continued to exercise its inherited levers of control in Ukraine while establishing new inroads into Ukrainian political and economic structures, the country's military and security services, media organizations, and other institutions.[13] The presence of oligarchs sympathetic to Moscow in Ukraine's political and economic system likely contributed to Ukraine's inability to pull away from Russia's influence further and more profoundly.[14] Ukraine has actively taken steps to distance itself from its eastern neighbor and limit Russia's ability to determine Ukraine's future. However, Kyiv's foundational disagreement over Moscow's rightful role in Ukraine has loomed over Russian-Ukrainian relations, fueling Ukrainian resistance movements over the past 30-plus years.

Orange Revolution (2004)

Ukraine's Orange Revolution in 2004 was a significant turning point in the country's democratization journey, its drive to assert independence from Moscow, and the development of the civilian-based resistance in the 2022 Russia-Ukraine War.[15] Importantly, several of the lead opposition figures responsible for coordinating the 2000–2001 movement calling for the resignation of President Leonid Kuchma "had earlier cut their organizational teeth on the mass protests of 1990" and were central figures in the

[12] Paul D'Anieri, *Ukraine and Russia: From Civilized Divorce to Uncivil War*, Cambridge University Press, 2019, p. 28.

[13] D'Anieri, 2019, p. 32.

[14] Stephen Grey, Tom Bergin, Sevgil Musaieva, and Roman Anin, "Putin's Allies Channelled Billions to Ukraine Oligarch," Reuters, November 26, 2014; Marek Dabrowski, "Ukraine's Oligarchs Are Bad for Democracy and Economic Reform," *Bruegel*, blog, October 3, 2017.

[15] Michael McFaul, "Ukraine Imports Democracy: External Influences on the Orange Revolution," *International Security*, Vol. 32, No. 2, Fall 2007, p. 49.

Orange Revolution.[16] Public confidence in Kuchma, Ukraine's second post-independence president, had waned following Kuchmagate (a major scandal in which Kuchma ordered the kidnapping of a journalist in 2000), and mass demonstrations broke out across Ukraine in 2004 after the revelation that the results of the presidential election had been doctored in favor of Kuchma and Kremlin-backed candidate Viktor Yanukovych.[17]

For 17 bitterly cold days, Ukrainians battled the elements to demonstrate their opposition to Yanukovych's attempts to hijack the presidency. Broadly speaking, the principles used to organize, fund, and sustain the Orange Revolution are echoed in the later Revolution of Dignity and in the context of resistance against Russia in the 2022 Russia-Ukraine War. For instance, locals opened their homes and businesses to house and feed demonstrators.[18] In 2004, donations from Ukrainians were used to fund the construction and heating of a tent city in the center of Kyiv and feed its residents.[19] The same phenomenon has unfolded since Russia's invasion in February 2022, with Ukrainian-based volunteerism and crowdfunding for everything from military kits to medical supplies (detailed below).[20]

The adoption and creative employment of modern communication technologies, particularly by young Ukrainians, were critical to the opposition movement in 2004. In the early 2000s, government censors had not yet attempted to regulate the use of websites as a means of information propagation. For this reason, the Ukrainian internet emerged as an important, unfettered tool for the Viktor Yushchenko campaign and the civilian-based

[16] Karatnycky, 2006, pp. 33–34.

[17] Anika Binnendijk, "Holding Fire: Security Force Allegiance During Nonviolent Uprisings," thesis, Fletcher School of Law and Diplomacy, Tufts University, August 2009.

[18] After tapes were leaked revealing that Kuchma had ordered the abduction of investigative journalist Heorhiy Gongadze, demonstrators took to the streets under the banner of "Ukraine without Kuchma." See Karatnycky, 2006, pp. 33–34, 40–41.

[19] Kateryna Panasiuk and Mykyta Vorobiov, "Charity Begins with Drones: Funding Ukraine's Resistance," Center for European Policy Analysis, September 28, 2022.

[20] Panasiuk and Vorobiov, 2022.

movement that supported him.[21] Demonstrators leveraged internet-based technologies to organize rallies, convey information, and provide updates on developments across Ukraine.[22] Youth groups, whose members were early adopters of what were then emerging technologies, organized cohorts to conduct door-to-door polls of Ukrainian households to identify which families could house activists—data that they compiled and communicated using cell phones and online databases.[23]

Likewise, Ukrainians leveraged web-based platforms and SMS messages to boost morale and build support for the opposition. For instance, "satire, jokes, and puns were often created in online chat rooms but were then distributed via mobile phone or live at protests to hundreds of thousands of people."[24] One example is the so-called egg incident. While on the campaign trail in late September 2004, Yanukovych was hit by an egg that a protester had lobbed at the candidate. Onlookers captured the politician's exaggerated response, including his dramatic fall to the ground, on video. Despite attempts by Yanukovych's team to bury it, the video surfaced. Ukrainians opposing Yanukovych sprang into action and used the incident as inspiration for satirical anti-Yanukovych messaging, such as jokes, online games, and skits.[25] Importantly, this tactic motivated those Ukrainians who would not have done so otherwise to "join the political conversation and [it] kindled the morale of existing revolutionaries on the Maidan."[26]

This incident parallels the messaging techniques employed by the resistance in Ukraine in the 2022 Russia-Ukraine War, including the use of

---

[21] Olena Prytula, "The Ukrainian Media Rebellion," in Ander Åslund and Michael McFaul, eds., *Revolution in Orange: The Origins of Ukraine's Democratic Breakthrough,* Carnegie Endowment for International Peace, 2006.

[22] Prytula, 2006, p. 110.

[23] Taras Kuzio, "Everyday Ukrainians and the Orange Revolution," in Ander Åslund and Michael McFaul eds., *Revolution in Orange: The Origins of Ukraine's Democratic Breakthrough,* Carnegie Endowment for International Peace, 2006, p. 81.

[24] Joshua Goldstein, "The Role of Digital Networked Technologies in the Ukrainian Orange Revolution," Berkman Center for Internet and Society, Harvard University, Research Publication No. 2007-14, December 2007, p. 12.

[25] Goldstein, 2007, pp. 11–12.

[26] Goldstein, 2007, pp. 11–12.

humor and appeals to emotion as a means of bolstering their morale and denigrating their adversaries. The early adoption of emerging technologies and their sophisticated use in service of Ukrainian resistance movements is a recurring pattern.

The Revolution of Dignity and the Evolution of the Ukrainian Resistance

Despite the scandal surrounding Yanukovych's bid for presidency that brought Ukrainians to the streets in 2004, he was nevertheless elected to that same office in 2010. In 2014, Yanukovych made the fateful decision to abandon Ukraine's anticipated Association Agreement with the European Union (EU) to pursue a Russo-centric foreign policy.[27] Ukrainians opposing the decision took to the streets in cities across the country, once again setting up tent communities in the harsh Ukrainian winter.[28] In response to the authorities' use of brutal tactics to disband the demonstrations, the resistance swelled in size and scope: Demonstrators began seizing government buildings around the country. Events came to a head in late February 2014, when demonstrators rushed the Presidential Palace, resulting in Yanukovych's flight from Kyiv and his removal from office by Ukraine's parliamentary body, the *Verkhovna Rada*.[29]

The approach of those demonstrators involved in the Revolution of Dignity, similar to the Orange Revolution before it, involved encampments in Kyiv and other cities around Ukraine and communal efforts to provide shelter and sustenance.[30] As characterized by historian Nadia Diuk, the resistance included "tents and field kitchens, and facilities for the people who were not merely episodically protesting but living there full-time."[31] But, as Diuk describes, the 2014 protests differed from previous grassroots efforts

[27] Nadia Diuk, "EUROMAIDAN: Ukraine's Self-Organizing Revolution," *World Affairs*, Vol. 176, No. 9, March–April 2014, pp. 9–14.

[28] Diuk, 2014, pp. 9–14.

[29] Yuriy Shveda and Joung Ho Park, "Ukraine's Revolution of Dignity: The Dynamics of Euromaidan," *Journal of Eurasian Studies*, Vol. 7, No. 1, January 2016.

[30] Diuk, 2014, p. 14.

[31] Diuk, 2014, p. 14.

in key ways. First, the resistance was not strictly driven by specific political figures, but a broader coalition of civic groups, student organizations, and others. Therefore, there was a recognition that Ukraine's future lay in the hands of its citizenry broadly, not in the hands of a select few.[32] This laid the groundwork for opposition during the 2022 Russia-Ukraine War, in which all segments of society take part in activism and there is a collective approach to resistance.

Second, new civil society organizations emerged in the wake of these events. Such groups include StopFake, which aims to raise awareness of and counter Russian disinformation. Although Ukraine has long possessed a robust civil society, the efforts and organizations that have developed since Russia's 2014 invasion have played a critical role in shaping contemporary civilian-based resistance efforts. The Atlantic Council has found that since Ukraine's 2014 Revolution of Dignity, "virtually every single meaningful reform in the country has been achieved thanks to the active participation of civil society."[33] For instance, the knowledge developed by grassroots civilian efforts to procure and distribute military kits for Ukrainian personnel in the Donbas starting in 2014 has likely proven valuable since Russia's invasion in 2022.[34]

---

[32] Diuk, 2014, pp. 14–15.

[33] Mykhailo Zhernakov, "Ukraine's Vibrant Civil Society Deserves Key Role in Post-War Transformation," Atlantic Council, July 19, 2022.

[34] Rosaria Puglisi, "A People's Army: Civil Society as a Security Actor in Post-Maidan Ukraine," working papers 15, Istituto Affari Internazionali, July 2015.

CHAPTER 3

Civilian-Based Resistance in Ukraine in the First Four Months of the 2022 War

Imposing Direct or Indirect Costs on an Occupying Force

The first proximate objective in Binnendijk and Kepe's framework for civil-based resistance is imposing direct or indirect costs on an occupying force. According to this framework, reaching this objective could change the occupier's calculations of "any benefits of an occupation against the sum of its costs" and "prompt a withdrawal, and achieve the outcome of national independence."[1] A wide range of activities may support this proximate objective, including sabotage and eliciting international sanctions. Imposing costs on Russia directly or via the international community has been a key priority for Ukraine. During a meeting with the Bucharest Nine on April 1, 2022, Ukrainian Minister of Foreign Affairs Dmytro Kuleba called on the participants to "weaken Russia in order to prevent it from achieving its goals."[2] During the first four months of the war in Ukraine, we observed

---

[1] Binnendijk and Kepe, 2021.

[2] The Bucharest Nine is an international organization founded in 2015 as a response to a perceived increase in Russian aggression. Its members are Bulgaria, the Czech Republic, Estonia, Hungary, Latvia, Lithuania, Poland, Romania, and Slovakia. See Izbirkom, "Хроніка Вторгнення Росія Затвердила Введення Військ На Схід України [The Chronicle of the Invasion of Russia Approved the Introduction of Troops to the East of Ukraine]," April 1, 2022.

a variety of civilian-based actions that appear to be designed to impose direct or indirect costs on the Russian forces.

Individuals and groups have leveraged their specialized technical and linguistic skills in support of Ukraine's fight against Russia's incursion. In Ukraine, cyber-based activities have taken on a much more prominent role than has been observed in other recent conflicts. It is the first major war involving large-scale cyber operations.[3] Cyber-based activities are advantageous because actors can be distributed globally. For example, the IT Army is a loosely bound group of amateurs and professionals hailing from Ukraine and elsewhere who volunteer their information technology (IT) skills in support of "hacking enemy resources, spreading viruses, creating phishing sites, and inflicting maximum damage on the [Russian] economy and other public spheres."[4] The group has allegedly worked in cooperation with the Ministry of Defense of Ukraine (MOD) officials. According to Oleksandr Bornyakov, Ukraine's Deputy Minister of Digital Transformation, the IT Army's lack of structure and absence of a formal chain of command make it more resilient.[5] It carries out cyber operations, targeting Russian infrastructure, government websites, Russian social media sites (such as Rossgram and RuTube), and various civilian service providers; cyber espionage; and defensive activities. Its activities are coordinated through a social media site.[6]

Likewise, Ukrainian supporters have also leveraged internet-based platforms as a means of disseminating messages about the realities of the war with Russia, in Russia. This tactic is a modern-day reincarnation of earlier

---

[3] James A. Lewis, "Cyber War and Ukraine," Center for Strategic & International Studies, June 2022; Maksym Butchenko, "Ukraine's Territorial Defence on a War Footing," International Centre for Defence and Security, April 13, 2022; Michael Schwirtz, Anton Troianovski, Yousur Al-Hlou, Masha Froliak, Adam Entous, and Thomas Gibbons-Neff, "Putin's War: The Inside Story of a Catastrophe," *New York Times*, December 16, 2022.

[4] IT Army of Ukraine, "IT ARMY of Ukraine," Telegram post, May 18, 2022, quoted in Stefan Soesanto, *The IT Army of Ukraine: Structure, Tasking, and Ecosystem*, Center for Security Studies, ETH Zürich, June 2022.

[5] Elise Labott, "'We Are the First in the World to Introduce This New Warfare': Ukraine's Digital Battle Against Russia," *Politico*, March 8, 2022.

[6] Soesanto, 2022.

leaflet campaigns with similar objectives. For instance, IT Army supporters have built a website that users can visit to send messages to Russians whose cell phone numbers and emails addresses these hackers have obtained.[7] Equally controversial is the Ukrainian use of the Clearview AI (artificial intelligence) facial recognition system—a company that has stirred international controversy because it harvests massive amounts of data without individuals' consent. Ukrainians have used this technology to identify deceased Russian soldiers and notify their next of kin via social media or messaging applications.[8]

Although these grassroots "hacktivist" activities might frustrate Russia on the margins and offer additional reconnaissance and psychological operations capabilities, cyber experts suggest that "sustained and systematic efforts" are required to impose significant costs on states such as Russia, where the public has less agency in affecting policymaking.[9] Notably, U.S. experts suggest that although the IT Army could engage in effective distributed denial-of-service (DDoS) attacks and defensive operations, hacktivist activities against Russian websites and civilian infrastructure—although highly visible—have not had a profound effect on Russia's military operations.[10] Moreover, some reports suggest that the lack of coordination among the many volunteer hacktivists, similar to volunteer contributions to kinetic military operations, may have undermined cyber-based operations that could have yielded potentially greater strategic effects.[11]

Benefiting from increasing digitization, civilians inside and outside Ukraine have also used open-source intelligence (OSINT) methodologies, such as analyzing commercial satellite data, air traffic transponder data, and geolocation data published on social media profiles; verifying images of attacks and civilian massacres; tracking the movement of Russian troops,

---

7 Danielle Kurtzleben, "Volunteer Hackers Form 'IT Army' to Help Ukraine Fight Russia," transcript of interview with Dina Temple-Raston, *NPR*, March 27, 2022.

8 Soesanto, June 2022.

9 See, for example, Lewis, 2022.

10 Lewis, 2022; and Matt Burgess, "Ukraine's Volunteer 'IT Army' Is Hacking in Uncharted Territory," *Wired*, February 27, 2022.

11 For example, a DDoS attack on a Russian website allegedly interrupted an ongoing website exploitation effort. See Kurtzleben, 2022.

and geolocating the yachts of Russian oligarchs.[12] Although this trend is not new—open-source data analysis was already used in the Iranian Green Revolution (2009), the Arab Spring (2010–2012), and the civil war in Syria (2011–), the scale of this so-called people's intelligence service in the Russia-Ukraine War (2022) is considerably larger.[13] Also, informal information networks between Ukrainian activists and their friends and acquaintances in the Russian border regions reportedly allowed Ukraine to receive information about the movements and location of Russian military equipment and troops.[14] Reports suggest that civilian-provided geolocation data may have been used by Ukrainian forces for targeting.[15] In fact, in March 2022, together with the Come Back Alive Foundation and other volunteers, the SBU created a web-platform to facilitate the civilian supply of geolocation information. This allowed civilian witnesses to input information about the location of Russian forces, their equipment, and any identification marks.[16]

Civilians have also attempted to impose costs on Russian forces through physical means. These tactics have included blocking Russian forces on the streets, protesting Russian troop presence, attacking Russian military vehicles, and avoiding cooperation with the Russian forces in occupied areas.[17] Particularly during the early stages of war, the media reported numerous instances wherein Ukrainian civilians attempted to block the movement of Russian military vehicles with their own bodies.[18] Other civil-

[12] Ivar Ekman and Per-Erik Nilsson, "Folkets Underrättelsetjänst—Öppna Källor 'OSINT' Och Ukraina [The People's Intelligence Service—Open Source "OSINT" and Ukraine]," Swedish Defence Research Agency, project number 1498503, June 2022.

[13] Ekman and Nilsson, 2022.

[14] Daza Sierra, 2022.

[15] Andrew Salerno-Garthwaite, "OSINT in Ukraine: Civilians in the Kill Chain and Information Space," *Global Defence Technology*, October 2022.

[16] LB.ua, "SSU, Return Alive Foundation and Volunteers Launch Joint Service," March 2, 2022.

[17] Jesse Lambert [@watchmans_way], "Russia sneaks into the country using civilian vehicles. This one clearly marked Z is bum rushed by dozens of citizens," Twitter post, March 1, 2022.

[18] "Ukrainian Resistance: Man Tries to Push Back Russian Tank - Video," *The Guardian*, February 26, 2022; ABC News, "Protesting Civilians Confront Russian

ians approached Russian soldiers to urge them to go home, speak about peace and the pointlessness of their actions, and blame Putin rather than soldiers—actions that have in some historic instances succeeded in undermining soldier loyalty to political leadership.[19] Although the effects of these types of grassroots efforts are difficult to calculate, reporting indicates that they have occasionally frustrated the progress of Russian forces.[20] In one reported case, a civilian protest in Kherson in March 2022 led some Russian trucks to turn around, thereby impeding the Russian advance, at least temporarily.[21]

Ukrainian military personnel have also credited local individuals for circumventing Russian attempts to jam Ukrainian communications, ensuring that Ukrainian forces are able to maintain an accurate picture of Russian movements.[22] For example, civilians with skills in drone manufacturing and operations have voluntarily produced, repaired, and operated low-cost spotter drones to detect Russian targets for the Ukrainian artillery.[23] Civilians seem to have also participated in identifying and moving abandoned

---

Military Vehicles in Ukraine," March 1, 2022; MilitaryLand.net [@Militarylandnet], "Civilians blocking the passage of another Russian column in #Melitopol #Ukraine #UkraineRussiaWar," Twitter post, March 1, 2022.

[19] A historical example of civilians successfully convincing invading soldiers to defect or leave is German citizens employing nonviolent measures to increase resentment and undermine the loyalty of the French forces in the Ruhr after World War I. During Ukraine's Orange Revolution in 2004, activists sought to motivate security forces officers to defect "by claiming that if the movement seized power, it would address officers' poor salaries and living conditions." See Maciej J. Bartkowski, "Countering Hybrid War: Civil Resistance as a National Defence Strategy," *OpenDemocracy*, May 12, 2015; and Sharon Erickson Nepstad, *Nonviolent Revolutions: Civil Resistance in the Late 20th Century*, Oxford University Press, 2011.

[20] Lambert, 2022.

[21] Chris Matthews, "Sending Putin's Invaders into Reverse: Defiant Ukrainians Chant 'Go Home' as They Force Back Two Russian Military Vehicles Marked with a 'Z' During Fearless Demonstration in Kherson," *Daily Mail*, March 20, 2022.

[22] Paul Sonne, Isabelle Khurshudyan, and Serhiy Morgunov, "Battle for Kyiv: Ukrainian Valor, Russian Blunders Combined to Save the Capital," *Washington Post*, August 24, 2022.

[23] Jason Bellini, "Ukrainian Civilians Use Drones to Help Fight Against Russian Forces," *Scripps News*, May 3, 2022.

Russian armored vehicles that can be wholly repurposed by Ukrainian forces or cannibalized for parts.[24]

Ukrainians have also sought ways to exploit divisions and reluctance among Russian soldiers to fight in Ukraine, drawing on their resentment, fear of brutal combat, and fear of dying.[25] These tactics involve summons from Ukrainian civilians and civilian government organizations to Russian soldiers to disengage from the war, President Volodymyr Zelenskyy's appeals to Russian mothers to prevent their sons from going to war, and the establishment of a dedicated Coordination Headquarters for the Treatment of Prisoners of War by the Ukrainian Cabinet of Ministers in March 2022 to signal that Ukraine was ready and willing to treat prisoners of war in accordance with international law.[26] Other related tactics have involved sending intimidating text messages or instructions for desertion protocols to Russian (and Belarusian) military personnel located in or near Ukraine.[27] Ukrainian civilians have capitalized on the confusion in the Russian forces caused by the lack of clear internal communication about the purpose of the war and the lack of trust between Russian soldiers and higher-level decisionmakers by confronting soldiers on the streets and demanding to know why they are in Ukraine. In an effort to exacerbate confusion, they have approached Russian soldiers to talk about the senselessness of war and ask them to go home.[28] In an attempt to undermine loyalty, Ukrainians have also sought to distinguish between the indifferent treatment of Russian soldiers by their commanders and their humane treatment in Ukraine even as prisoners of war by publishing videos of Russian soldiers being fed and allowed

---

[24] Hromadske Int. [@Hromadske], "Grand Theft Russian Military Vehicles: Ukraine," Twitter post, March 6, 2022.

[25] Pjotr Sauer, "'They Were Furious': The Russian Soldiers Refusing to Fight in Ukraine," *The Guardian*, May 12, 2022.

[26] Ukrinform, "Coordination Headquarters for Treatment of Prisoners of War Established in Ukraine," March 12, 2022.

[27] Michael Weiss, "Inside Ukraine's Psyops on Russian and Belarusian Soldiers," *New Lines Magazine*, March 29, 2022.

[28] MilitaryLand.net, 2022.

to call home after capture.[29] Together with videos of Russian troops being interrogated, however, these videos created a backlash. Ukraine has been accused of violating Article 13 of the Geneva Convention, which requires that prisoners of war be protected from "insults and public curiosity."[30]

The authorities in Kyiv have also aimed at disrupting the deployment and movement of Russian forces and, by extension, imposing costs on Russia. The Ukrainian government sought to appeal to the more commercially minded soldiers by promising five million Russian rubles (approximately $45,000) in cryptocurrency and amnesty to Russian soldiers in exchange for their surrender.[31]

Lastly, Ukrainian activities aimed at inflicting political and economic costs—such as politically isolating Russia from the world and lobbying for international sanctions—have also been prominent. For instance, the Ukrainian government has pushed for formal international recognition of Russia as a terrorist state.[32] More-indirect means of imposing costs on the Russian forces include financial or in-kind donations to the Ukrainian military forces or volunteering time to support the Ukrainian military effort. There are numerous initiatives aimed at collecting financial or in-kind support for the Ukrainian Armed Forces (including an online donation site started as an initiative by President Zelenskyy), organized by individuals or

---

[29] Timofei Rozhanskiy, "Why Russian Soldiers Are Refusing to Fight in the War on Ukraine," *Radio Free Europe/Radio Liberty*, July 20, 2022; Richard Sisk, "Captured Russian Troops Call Home While Filmed by Ukrainian Officials, Raising Geneva Convention Questions," Military.com, March 1, 2022.

[30] United Nations, *Geneva Convention Relative to the Treatment of Prisoners of War*, adopted August 12, 1949, by the Diplomatic Conference for the Establishment of International Conventions for the Protection of Victims of War, held in Geneva from April 21 to August 12, 1949.

[31] Lambert, 2022; and Pradipta Mukherjee, "Russian Soldiers Offered Bitcoin in Exchange for White Flag," *Yahoo Finance*, February 27, 2022.

[32] Volodymyr Zelenskyy, "Recognition of Russia as Terrorist State Needed Not as Political Gesture, but as Effective Defense of Free World—Address of President of Ukraine," Office of the President of Ukraine, July 30, 2022.

groups of individuals by donating international sports trophies or personal savings to fundraisers.[33]

Overall, it is difficult to assess the extent to which civilian-based resistance actions specifically have influenced the outcomes of the conflict. Instead, a comprehensive assessment should be conducted after the end of the war. However, we can say that Ukrainian civilians have engaged in activities that may have increased personnel and materiel requirements for Russian forces, altered at least some Russian soldiers' and some segments of Russian society's support for the war, and increased economic and political costs for Russia. By extension, it is possible that these activities have altered the balance between Ukrainian and Russian military capabilities.

Because the war in Ukraine has involved the evolution and novel application of civilian-based initiatives, it has raised new questions for policymakers and scholars. For instance, international laws about cyberspace, the involvement of foreign companies or individuals in direct offensive resistance, and the use of certain new technologies, such as biometric databases and AI recognition, all require further examination.[34] One author suggests that Ukraine's internet-based activities have already "collapsed entire pillars of existing legal frameworks regarding norms and rules for state behaviour in space and has taken apart the illusion of separating the defense of Ukraine from Ukrainian companies and citizen[s] living abroad."[35]

Securing External Support

Resistance literature and historical precedents suggest that the ability to attract external support is key to maintaining armed and unarmed resistance. External support provides an avenue for supplying resources to the

[33] Government of Ukraine, "Donate to Ukraine's Defenders," webpage, undated-b; Ministry of Youth and Sports of Ukraine, "Українські Спортсмени Жертвують Своїми Найціннішими Нагородами Для Захисту України Від Російської Агресії Та Змінюють Спортивну Форму На Військову [Ukrainian Athletes Donate Their Most Valuable Awards to Protect Ukraine from Russian Aggression and Change their Sports Uniforms to Military Uniforms]," April 6, 2022.

[34] Soesanto, 2022.

[35] Soesanto, 2022.

armed and the unarmed resistance and defense efforts. External political and public support can raise awareness of the occupation and help achieve international actions, such as sanctions. Moreover, international pressure may also be used to influence the adversary's political and military aims or willingness to engage in negotiations or withdraw forces, and therefore has coercive value.[36] Though initially threatened, the capital city of Kyiv and its resident state functions have been protected, and its civilian population has generally maintained its capacity to impose costs on Russian occupying forces. Therefore, most of the country continues to function under Ukraine's democratically elected government and has not been forced to build resistance activities from exile. Nevertheless, in seeking external support, Ukrainian actors have combined historical methods of civilian-based resistance with more-novel approaches, including the use of new technologies and leveraging the information environment. Ukraine's solicitations for external support included not only seeking material and political support for its resistance activities but also requesting that other external actors impose costs on Russia through sanctions and by cutting business and cultural links.

During the first four months of the war, securing continuous support from external actors—including foreign governments, international organizations, enterprises, and individuals—was one of the key elements of Ukrainian civilian-based activities. The initial shock of war in Europe led numerous other governments, nongovernmental organizations, and individual actors to volunteer or donate nonmilitary and military assistance to Ukraine. Ukrainian government actors, enterprises, and individuals engaged in a wide variety of activities to solicit and maintain external political, financial, and material support for Ukraine and help impose economic and diplomatic costs on Russia. In other words, although Ukrainian government actors worked to secure external political and military support for Ukraine's war efforts, there were also very active grassroots efforts to solicit external support for Ukrainian civilians and the Ukrainian defense efforts.

Since February 2022, President Zelenskyy has become the most-visible lobbyist on behalf of Ukraine. He has become the embodiment of Ukraine's

[36] Binnendijk and Kepe, 2021.

resistance to Russia's occupation through his speeches that address multiple audiences and "create urgency, tension, and immediacy."[37] He seems to have captured the admiration of international and domestic audiences through his carefully crafted rhetoric, which emphasizes defiance toward the Russian invasion, and his image as "a man for the people, of the people."[38]

Binnendijk and Kepe's analytical framework includes the actions of the state's diplomatic service. The Ukrainian diplomatic service, its government, and its population have been active in engaging with external nonmilitary support.[39] Ukraine's diplomatic service has been lobbying the international community to recognize the Russian Federation as a terrorist state and violator of international laws and human rights and to formally acknowledge that Russia has not met its obligations as a member of the United Nations (UN) and the Organization for Security and Co-Operation in Europe (OSCE). The Ukrainian diplomatic service has also used international organizations to keep the war in Ukraine on the international agenda and solicit political and practical support from the international community.[40] As a result, OSCE member countries invoked the OSCE Moscow Mechanism twice since February 2022. Established in 1991 to allow the OSCE to send expert missions to help countries resolve issues with a specific human dimension, this tool has been used to investigate and document the Russian Armed Forces' violations of international human rights law and human rights abuses in Ukraine.[41] Although Ukraine's calls for Russia to

---

[37] Kvartal 95, "About Us," webpage, undated; and Nomi Claire Lazar, "Need an Expert? War in Ukraine: President Volodymyr Zelenskyy's Rhetoric," University of Ottawa, March 22, 2022.

[38] Ajnesh Prasad, "Volodymyr Zelensky's Appeal Lies in His Service to Ukrainians Above All Else," *The Conversation*, March 2, 2022; and Paul Adams, "'Shame on you': How President Zelenskyy Uses Speeches to Get What He Needs," *BBC*, March 24, 2022.

[39] At the time of writing, it is difficult to confirm the extent to which the Ukrainian central government has made soliciting external support the focus of its messaging, as the Ukrainian government has been tight-lipped about its operations.

[40] Ministry of Foreign Affairs of Ukraine, "Організація З Безпеки І Співробітництва В Європі (ОБСЄ)," ["Organization for Security and Cooperation in Europe (OSCE)]," July 20, 2022.

[41] Organization for Security and Co-Operation in Europe, "Moscow Mechanism," December 1, 1991; and Ned Price, "Invocation of the OSCE Moscow Mechanism to

be expelled from the UN Security Council have not been successful to date, the UN General Assembly voted to suspend Russian from the UN Human Rights Council in April 2022.[42]

Ukraine's business associations and enterprises have sought innovative means for attracting support from abroad. Examples include creating an online platform that facilitates direct donations from around the world to local Ukrainian businesses, which has allowed them to offer free services to Ukrainians; this has helped compensate for Ukraine's shrinking economy and purchasing power in addition to the shrinking reserves of Ukraine's small businesses.[43] At the same time, numerous external assistance activities seem to be motivated by external interests. The Polish Investment and Trade Agency supports small and medium-size businesses by offering them office space and the ability to carry out their work from Poland to support Ukraine's long-term ability to recover.[44] Meanwhile, lobbyists acting on behalf of the Ukrainian government and industry in the United States have shifted their focus away from energy: specifically, stopping the development of Russia's Nord Stream 2 pipeline. Although the work of these lobbyists was affected by the entanglement of Ukraine in U.S. domestic politics during the Trump administration, they now lobby to ensure military support and sanctions on Russia.[45] Public pressure—from both consumers and investors—has also targeted Western companies that were doing business in Russia. This resulted in almost 1,000 companies pledging to exit or

---

Investigate Mounting Reports of Human Rights Abuses and International Humanitarian Law Violations by Russia in Ukraine," press statement, U.S. Department of State, June 3, 2022.

[42] "Zelenskyy Tells UN That Russia Must Be Expelled from Security Council," *France 24*, April 5, 2022; Michelle Nichols, "U.N. Suspends Russia from Human Rights Body, Moscow Then Quits," Reuters, April 7, 2022.

[43] Maria Prus, "Як Підтримка Малих Бізнесів Може Допомогти Швидшому Відродженню Економіки України [How Supporting Small Businesses Can Help Speed Up the Revival of Ukraine's Economy]," *Ukrainian VOA*, May 4, 2022.

[44] Polish Investment and Trade Agency, "The Polish Investment and Trade Agency (PAIH) Provides Free Office Space for Ukrainian Companies," March 1, 2022.

[45] Julie Bykowicz and Vivian Salama, "Ukraine Lobbyists in U.S. Pivot from Pipeline to War Aid and Sanctions," *Wall Street Journal*, March 23, 2022.

already exiting or cutting down their operations in Russia by the beginning of June 2022.[46]

Ukraine's substantial diaspora communities abroad have also played a significant role. The Ministry of Foreign Affairs of Ukraine reports that the largest Ukrainian communities abroad (besides Russia) reside in Canada and the United States, with approximately 1.2 million and 0.9 million Ukrainians, respectively. According to official statistics, more than 100,000 Ukrainians reside in Brazil, Kazakhstan, Moldova, Argentina, Belarus, and Germany, with additional considerable communities in other European countries, Australia, and South America.[47] Popular actions include rallies and protests on the streets and on social media, using social media to counter Russian disinformation and solicit the support of foreign audiences, directly addressing foreign decisionmakers, using personal networks to solicit and deliver help to people in Ukraine and Ukrainian refugees, and using Ukrainian cultural heritage to ensure popular support for Ukraine and gather funding for humanitarian aid. For example, the Ukrainian Congress Committee of America, Illinois Division, has set up a relief fund for Ukraine, bought medical equipment, and organized numerous events in support of Ukraine in Chicago.[48]

Although it is difficult to assess the extent to which external support resulted from specific Ukrainian activities versus self-motivated volunteerism, Ukraine was on the whole extremely successful in attracting external support from governments, enterprises, and private individuals in the first months of the war. Moreover, it did so rapidly. By June 2022, the EU alone had allocated €4.1 billion in assistance to Ukraine, in addition to military assistance, support to Ukrainian refugees, and several rounds of

[46] Ciara Linnane, "Companies That Exited Russia After Its Invasion of Ukraine Are Being Rewarded with Outsize Stock-Market Returns, Yale Study Finds—and Those That Stayed Are Not," *Market Watch*, June 8, 2022.

[47] Ministry of Foreign Affairs of Ukraine, "Ukrainians Worldwide," December 18, 2019.

[48] Ukrainian Congress Committee of America, Illinois Division, "Letter to the Honorable Antony Blinken," Facebook post, August 2, 2022; RAND researcher's review of Ukrainian Congress Committee of America, Illinois Division, Facebook page, undated.

sanctions against Russia.[49] In July 2022, the EU announced the allocation of the first billion of the additional €9 billion financial support package for Ukraine.[50] By July 9, the United States had committed over $1.28 billion in humanitarian assistance to Ukraine. Together with U.S. security assistance efforts, these funds made the United States the largest single-country donor of humanitarian assistance to Ukraine.[51] These are only some of the external donors. Nonmilitary (financial or humanitarian) support to Ukraine has been provided bilaterally or as part of multinational initiatives by other countries as well.

At the time of this writing, it remains unclear how long key political and economic supporters of Ukraine in Europe and North America will be able to maintain existing levels of support while battling domestic, political, and economic problems irrespective of the activities that Ukrainian actors may use to call for support. The war in Ukraine also calls into question the role of multinational private enterprises and companies (that are not part of the defense industry) in diplomacy and international armed conflicts.[52]

Denying an Occupier's Political and Economic Consolidation

A country under invasion may succeed in reasserting the control and legitimacy of its government, preserving its political and economic structures, and creating the foundation for an easier transition to postwar governance by preventing the occupier's ability to consolidate political and economic power over it.[53] Broadly speaking, the core activities aimed at denying Rus-

---

[49] Kristin Archick, "Russia's Invasion of Ukraine: European Union Responses and Implications for U.S.-EU Relations," Congressional Research Service, updated July 28, 2022.

[50] Archick, 2022.

[51] Antony J. Blinken, "Additional Humanitarian Assistance for the People of Ukraine," U.S. Department of State, July 9, 2022b.

[52] Laura Hautala, "Big Tech's Support for Ukraine Recasts Industry's Global Role," *CNET*, March 1, 2022.

[53] Binnendijk and Kepe, 2021.

sia's political and economic consolidation in Ukraine have included efforts to maintain the status and visibility of the democratically elected leadership of Ukraine, thwart Russian-organized referenda in Russian-occupied regions, and deny Russia the ability to control Ukrainian energy and trade. Most of the efforts that we observed were government-led or government-coordinated actions. Most civilian efforts seem to have been aimed at protesting against the Russian invasion inside and outside Ukraine and any preparations for Russian-organized referenda about the future of certain regions of Ukraine.[54]

Ukrainian efforts to preserve the continuity of government have focused on preserving Zelenskyy's public image as a symbol of Ukraine's resistance and the "guarantor of state sovereignty [and] territorial integrity."[55] Zelenskyy's decision to remain in Kyiv, particularly during the first month of the war when Russia tried to seize Kyiv, likely "contributed to Ukraine's ability to maintain and increase public support for Ukrainian national resistance in the country and abroad."[56] When the civilian resistance was still burgeoning, the president's role in the first months of the war—notably, his public demonstrations of his commitment to his post and Ukrainian sovereignty—likely boosted morale and served as a rallying point for resistance. His government's unified voice was critical in providing clarity on the situation in Ukraine and reaffirming Ukraine's political stance.[57]

Given Zelenskyy's central role as the symbol of Ukraine's resistance, the prospect of his capture or death poses considerable risks and could "deal a psychological blow to the morale of Ukrainians and diminish their will to resist."[58] This was especially true during the first months of war when it was particularly important to have a single unified center of resistance

[54] "Kherson: How is Russia Imposing its Rule in Occupied Ukraine?" *BBC*, May 11, 2022.

[55] Government of Ukraine, "Конституція України—Розділ V [Constitution of Ukraine—Chapter V]," undated-a.

[56] Marta Kepe and Anika Binnendijk, "What Is Continuity of Government, and Why Does It Matter for Ukraine's National Resistance?" *RAND Blog*, March 17, 2022.

[57] William Courtney and Khrystyna Holynska, "Continuity of Government in Ukraine," *RAND Blog*, February 25, 2022.

[58] Courtney and Holynska, 2022.

during a period when resistance activities were only emerging.[59] Although the U.S. Secretary of State has confirmed that Ukraine has plans for continuity of government, the details of such plans are unknown.[60] However, at least some parts of the line of succession of political leadership are known. In the case the powers of the president are terminated early, the Chairman of the Verkhovna Rada assumes the duties (albeit slightly limited duties) of the president until elections can be held.[61] In the event of a decapitation of the government, a clear and openly communicated line of presidential succession beyond the very top few officials could "expand the resilience of Ukraine's democratic governance beyond the vulnerability of any one individual or group."[62] Moreover, it could "protect a legitimate source of operational leadership and morale for the remaining resistance actors" and "help to ensure that Ukrainian, rather than Russian-chosen leaders gain rapid acceptance from the Ukrainian citizenry, and from the global audience, should a full occupation scenario unfold."[63]

Ukraine severed its diplomatic relations with Russia following the February 24, 2022, invasion and suspended the functions of its embassy in Moscow and consular services across Russia.[64] It continues to maintain a formal phone number and an email address for cases in which there is a life or death threat to Ukrainian citizens.[65] While closing the diplomatic representations in Russia may affect Ukraine's ability to liaise with the Ukrainians who were deported to Russia during the war or the large Ukrainian community there, Ukrainian nonprofit organizations such as *Krym SOS* have been educating the district administrations of the affected Ukrainian

[59] Courtney and Holynska, 2022.

[60] Antony J. Blinken, "Secretary Blinken with Margaret Brennan of CBS News," U.S. Department of State, March 6, 2022a.

[61] Government of Ukraine, undated-a.

[62] Kepe and Binnendijk, 2022.

[63] Kepe and Binnendijk, 2022.

[64] Evropeiska Pravda, "МЗС Повідомляє Про Припинення Роботи Усіх Консульств України В Росії [The Ministry Of Foreign Affairs Announces The Suspension Of All Ukrainian Consulates In Russia]," March 13, 2022.

[65] Embassy of Ukraine in the Russian Federation, homepage, undated.

regions (i.e., Kirovohrad, Mariupol, and Bakhmut) about how Ukrainians can return to Ukraine with expired or missing passports.[66]

Ukrainian organizations have sought to undermine potential Russian-organized regional referenda by suspending the State Voter Register on the day of the invasion to limit Russian access to voter lists.[67] Ukrainian security services have warned the population about Russian campaigns that "are intended to create an illusions that the 'vote' results voiced by the Russians reflect the will of local residents."[68] However, other external factors may influence whether Russia would be able to claim enough participants in a referendum; it has been reported that Russians might base their voting lists on the information gathered when distributing humanitarian aid, through forced passportization, and by accessing region-based electoral databases, or they might trade humanitarian aid or money for participation in a referendum.[69] In March 2022, the Ukrainian government created a legal mechanism for banning the activities of political parties that are pro-Russian, are accused of collaborationism, or undermine state sovereignty. As a result, Ukraine suspended the activities of 11 parties.[70] Together with the consolidation of Ukrainian TV platforms under a state platform, these measures have caused concerns in the international community about the use of domestic tools for political rather than national security purposes.[71] In addition to Ukraine's governmental actions, media have reported that

[66] Crimea SOS, "Кримsos: В Донецькій Області Розповіли Про Шляхи Повернення Українців З Рф [CrimeaSOS: In the Donetsk Region, They Talked About Ways to Return Ukrainians from the Russian Federation]," April 22, 2022.

[67] Olga Aivazovska, "Pseudo-Referendums in the Invaded Territories of Ukraine: What Russia Plans to Do and How to Disable Them," Opora, August 10, 2022.

[68] Aivazovska, 2022.

[69] Aivazovska, 2022.

[70] Opora, "Як Львівський Суд Забороняв Проросійські Політичні Партії. Аналіз 16 Судових Процесів [How the Lviv Court Banned Pro-Russian Political Parties. Analysis of 16 Trials]," August 17, 2022.

[71] Nick Fenton and Andrew Lohsen, "Corruption and Private Sector Investment in Ukraine's Reconstruction," Center for Strategic and International Studies, November 8, 2022; "Why Ukraine's Richest Oligarch Gave Up His Media Empire," *Yahoo News*, July 13, 2022.

civilians have protested against the installed pro-Russian government of the occupied Kherson region and its declared aim to reincorporate the region into Russia.[72]

Popular protests in Ukraine and by the Ukrainian diaspora erupted with the invasion. Ukrainians residing abroad and other supporters of Ukraine organized protests at Russian embassies, along with solidarity rallies, demonstrations, and prayer services.[73] These activities exhibited widespread outrage in response to the Russian invasion. It is possible, though difficult to say with any certainty, that these early events could have affected popular support for the Ukrainian cause both within Ukraine and beyond its borders and might have affected Russia's ability to consolidate its political aims in Ukraine.

The 2022 Russia-Ukraine War illustrates the importance of ensuring state food and energy reserves. The Ukrainian government lacked the food reserves to address at least the initial shortages.[74] In April 2022, it adopted an order "On the Approval of the Plan of Measures to Ensure Food Security in the Conditions of Martial Law" and had to scramble to buy products included in the list of critical goods for consumption (a 12.5-kilogram set of products that can be stored for a longer period, such as cereals, oil, canned goods, dairy and meat products, potatoes, crackers, and cookies).[75] These goods were then distributed and used to create a reserve for potential blockades, while farmers gave away their produce for free to support the population.[76] Ukrainian citizens self-organized to deliver food, medicine,

[72] "Kherson: How Is Russia Imposing its Rule in Occupied Ukraine?" 2022.

[73] Ukrainian Congress Committee of America, "Senate Ukraine Caucus," webpage, undated; and Legistorm, "Senate Ukraine Caucus," webpage, undated.

[74] Ministry of Economy of Ukraine, "Аудит 'Держрезерву' Підтвердив Тотальне Розкрадання Запасів І Мільйонні Збитки, ["The Audit of 'State Reserve' Confirmed the Total Theft of Stocks and Millions of Losses]," November 5, 2020.

[75] Cabinet of Ministers of Ukraine, "Розпорядження 'Про Затвердження Плану Заходів Забезпечення Продовольчої Безпеки В Умовах Воєнного Стану' [Order 'On the Approval of the Plan of Measures to Ensure Food Security in the Conditions of Martial Law']," No. 327, April 29, 2022.

[76] Ministry of Agrarian Policy and Food of Ukraine, "Тарас Висоцький: Україна сформувала продовольчі резерви для своїх громадян [Taras Vysotsky: Ukraine Has Formed Food Reserves for Its Citizens]," July 22, 2022.

and other emergency supplies to those in need through high-rise buildings, libraries, or other local community groups.[77] Similarly, Russia's invasion of Ukraine has shown the importance of adequate energy reserves and the benefits of an energy network that is integrated with friendly countries. Weakened by preexisting energy sector issues,[78] Ukraine and the international community have had to accelerate Ukraine's integration into the European energy networks to rapidly diversify its energy imports and exports.[79]

The homes of many Ukrainians who were unable or unwilling to relocate have sustained considerable physical damage. In addition to presenting a practical challenge, this wreckage serves as a visible scar of Russian attempts to devastate Ukrainians' will to resist. Again, Ukrainians established volunteer organizations that took on the role of coordinating cleanup efforts to clear the rubble from people's homes and businesses.[80] Civilian authorities have also been quick to repair infrastructure in cities that have sustained damage from Russian missile strikes. In several cases, roads have been cleared of debris and repaired within the span of a day. Officials have publicly boasted about their swift recovery efforts, casting the country's physical structures as emblems of Ukraine's enduring willpower.[81] In the

[77] Dan Vaccaro, "'Islands of Hope': Ukrainian Libraries Respond to Russian Invasion in Surprising and Heartbreaking Ways—Here's How," *Berkeley Library, University of California*, April 7, 2022; Oleksandr Lapshin, "Як мешканці одного будинку стали волонтерити разом. Історії з Харкова? [How Residents of the Same House Began to Volunteer Together. Stories from Kharkiv]," *Ukrainska Pravda*, April 17, 2022.

[78] Volodymyr Omelchenko, "Базові причини енергетичної кризи в ОЗП 2021–2022pp [Basic Causes of the Energy Crisis in the 2021-2022 OZP.]," Razumkov Centre, November 24, 2021.

[79] U.S. Department of Energy, "Ukraine Launches Electricity Exports to European Union with Support from the U.S. Department of Energy," July 2, 2022; Andrian Prokip, "Stability Against All Odds: Ukrainian Energy During the First Weeks of the War," *Wilson Center*, blog, March 28, 2022.

[80] Salma Abdelaziz, "As Russian Attacks Continue to Pound Ukraine, Volunteers Are Rebuilding Kyiv's Suburbs," *CNN*, June 13, 2022.

[81] "Ukraine Repairs Roads Smashed by Russian Missiles in Kyiv, Dnipro Overnight—PHOTOS," *Euromaidan Press*, October 11, 2022; Kyrylo Tymoshenko, "Бульвар Тараса Шевченка [Taras Shevchenko Boulevard]," Telegram post, October 11, 2022; Defense of Ukraine [@DefenceU], "We will rebuild everything that was destroyed. Our spirit will remain unbroken," Twitter post, November 3, 2022.

immediate aftermath of Russia's October 10, 2022, assault on Kyiv, which damaged the city's Glass Bridge, for example, Kyiv's mayor Vitali Klitschko reassured Ukrainians,

> the lighting [of the bridge] will be repaired. We are ordering the glass. It will take up to three weeks to make it. And soon our renovated bridge will once again delight the people of Kyiv and visitors to the capital. Because it is now not only a symbol of Kyiv for tourists, but also a symbol of the indomitability and stability of Ukraine.[82]

Furthermore, denying Russia the ability to control Ukraine's grain exports has become not only a matter of resisting Russia's control over Ukrainian sea lines of communication but also has highlighted the cascading multinational implications of military action and resilience-building and the role of external support to achieve these aims. In this case, the UN brokered a deal that allowed Ukraine to export grain from its Black Sea port that had been blocked by Russia.[83]

Reducing an Occupier's Capacity for Repression

There are two ways in which the fourth proximate objective identified by the Binnendijk and Kepe framework—activities aimed at reducing an occupier's capacity for repression—can advance a resistance movement's progress toward ousting an occupying force. First, and perhaps most evidently, the resistance movement can shield civilian populations from violence and abuse at the hands of the oppressor. This is a significant outcome in its own right. Second, activities designed to reduce an occupier's capacity for repression can also serve a resistance movement's broader objective of expelling the occupying force. Binnendijk and Kepe explain this relationship, noting that "exposing acts of repression can create new costs associated

[82] Olena Roshchina, "Kyiv Traffic Is Moving Again at Strike Point Near Shevchenko Park and Repair Work Begins on 'Klychko Bridge,'" *Ukrainska Pravda*, October 11, 2022.

[83] Bill Tomson, "Negotiations Intensify to Save Deal on Ukraine Grain Exports," *Agri-Pulse*, November 14, 2022.

with repression that could damage the international or domestic position of the occupier."[84] In other words, by identifying and broadcasting the repression inflicted on civilian populations, resistance movements can engender sympathy for their position on the international stage. In turn, doing so can convince external partners and allies to "impose costs on the occupier or to provide the resistance with other forms of support."[85]

We see evidence of Ukraine's civilian-based resistance working to reduce Russia's capacity for repression in service of both ends outlined above. Efforts to shield the civilian population from Russian repression have taken several forms. Most directly, volunteers have removed civilians from immediate threats. Since the initial invasion on February 24, 2022, many Ukrainian volunteers have risked their personal safety to evacuate vulnerable civilians (such as the elderly and disabled) from territories targeted by Russian shelling, Russian-held areas, and other spaces where civilian lives are under threat.[86] Unfortunately, the underprepared civilian protection and emergency response system may have contributed to the need for civilian volunteers to support and protect their fellow citizens. Prior to February 2022, Ukrainian sources reported that only 10 percent of the country's 21,000 bomb shelters were ready for an emergency and that state reserves of food, grain, and nonfood items (such as gasoline, diesel fuel, and coal) were in critical condition.[87] Kyiv's emergency evacuation plan was approved on February 12, 2022, only 12 days before the invasion.[88]

[84] Binnendijk and Kepe, 2021, p. 24.

[85] Binnendijk and Kepe, 2021, pp. 24–25.

[86] Natalie Thomas, "Volunteers Evacuate the Elderly from Ukraine's Bakhmut, Fearing Russian Advance," Reuters, June 28, 2022; Elena Becatoros, "Ukraine: Drivers Risk All to Bring Aid, Help Civilians Flee," Associated Press, June 10, 2022.

[87] Boris Sachalko and Andrey Shurin, "Офисы Под Землей И Указатели В Никуда. В Каком Состоянии Находятся Бомбоубежища В Украине [Offices Underground and Signs to Nowhere. In What Condition Are the Bomb Shelters in Ukraine]," December 12, 2021; Ukrinform, "У Держрезерві Заявляють Про Критично Низькі Запаси Зерна [The State Reserve Declares Critically Low Grain Reserves]," September 20, 2021; Ministry of Economy of Ukraine, 2020.

[88] New Format, "У Києві Затвердили План Евакуації Населення На Випадок Надзвичайної Ситуації [In Kyiv, an Emergency Evacuation Plan Was Approved]," February 12, 2022.

The level of coordination among volunteers offering evacuation services has varied. In some cases, drivers run their own one-person operations for profit; others operate for free. Some volunteers also organize into informal grassroots networks in which more-seasoned individuals share best practices with newcomers.[89] Others evacuate vulnerable civilians as part of formal initiatives organized by institutions such as Vostok SOS, a Ukrainian nongovernmental organization responsible for transporting over 25,000 civilians to safety.[90] These evacuation journeys often serve dual purposes: Once drivers and evacuees arrive safely at their destinations, often in Western Ukraine, many volunteers collect humanitarian supplies before heading back to the front lines, where they deliver aid to beleaguered populations in Russian-controlled territory.[91]

Once evacuees arrive in Western Ukraine, Kyiv, or other unoccupied territories in the country, reports indicate that local volunteers—individuals and organizations alike—provide internally displaced people with temporary shelter, food, supplies, and services.[92] Among other things, these services include psychological counseling and access to education.[93] Similar to many of the other resistance activities outlined in this report, these efforts appear to exist on a spectrum that includes individuals opening their personal homes to the displaced, private businesses devoting part or all of their infrastructure and services to the cause, and local governments converting public buildings into makeshift shelters.[94]

Once the initial tide of refugees flooding in from war-torn areas was stemmed, these same formal and informal resistance networks mobilized

[89] Becatoros, 2022.

[90] Kostyantyn Chernichkin and Anton Protsiuk, "Vostok SOS Helps Evacuate over 23,000 People Fleeing Russia's War, Continues Its Mission," *Kyiv Independent*, July 25, 2022; and Vostok SOS, "About Us," webpage, undated.

[91] Becatoros, 2022.

[92] Patrick J. McConnell, "In Ukraine's Lviv, Help—and Maybe Hope—for the Legions Displaced by War," *Los Angeles Times*, March 24, 2022.

[93] Voices of Children, "The Results of the Fund Work During of the Full-Scale War in Ukraine," September 4, 2022.

[94] See, for instance, Anya Kamenetz, "Shelter.Lviv Started on Instagram. It's Now Helped House 4,000 Women and Children," *NPR*, May 17, 2022.

to provide more sustainable short-term housing. The mayor and local government of the Western Ukrainian city of Lviv, for instance, took steps to build long-term housing communities for the city's 50,000 new residents. According to the city's head architect, these will likely take shape as "apartment blocks of five to seven stories . . . that would combine beauty with utility" and will be replete with bomb shelters and rooms reinforced against chemical weapons attacks, per Lviv's new building codes.[95] In the meantime, the city has erected prefabricated shipping container communities to house some of the city's new residents.[96]

All the activities outlined above, and numerous others like them, are critical elements in the efforts of the Ukrainian civilian-based resistance to reduce Russia's capacity for repressing noncombatants. By physically removing vulnerable civilians from ongoing hostilities, protecting those who cannot leave, and furnishing those stranded in perilous situations with aid, the Ukrainian resistance is directly limiting Russia's capacity to inflict pain on these populations, if even marginally. Such efforts also have the added benefit of bolstering Ukrainians' will to resist. In the areas under Russian occupation, however, the threat and fear of Russian repressions may have subdued civilian-based protest actions.[97]

Additionally, the resistance in Ukraine has vociferously publicized atrocities committed against noncombatants to further tarnish Moscow's image and its legitimacy in the eyes of foreign publics and decision-makers. To this end, Ukrainian officials have relied on traditional and social media to convey the atrocities that Ukrainian civilians have endured at the hands of the Russian military. Early in the conflict, for instance, Ukrainian President Zelenskyy embarked on a virtual speaking tour during which he addressed parliamentary bodies of partner and ally states to request assistance. Rather than rely on the power of his voice alone, Zelenskyy chose to

[95] Jane Arraf, "To House Refugees, Lviv Wants to Make Beautiful Buildings That Last," *New York Times*, May 31, 2022, updated June 2, 2022.

[96] Bloomberg Quicktake, "Life in a Shipping Container: Ukraine's Refugees in Lviv," video, September 19, 2022.

[97] According to Felipe Daza Sierra's review of resistance activities in Ukraine, public demonstrations in the occupied territories significantly decreased starting in April 2022. See Daza Sierra, 2022.

include an emotional video with grisly imagery exposing Ukraine's civilian casualties and human suffering and use emotionally powerful rhetoric.[98] He closed the speech by addressing President Joseph Biden directly: "I wish you to be the leader of the world. Being the leader of the world means to be the leader of peace," he said, in English.[99]

Hours after viewing the video and its accompanying remarks, Biden announced from the White House that the United States would provide an additional $800 million in assistance to Ukraine.[100] Legislators—many of whom were reportedly brought to tears by the video—rallied around Zelenskyy's cause, publicly urging for additional military aid to Ukraine. While we cannot be sure about the specific causal driver, the timing suggests that Ukraine's visual demonstration of Russia's oppression might have contributed to Biden's decision.

We see Ukrainian civilian officials taking a similar approach following the liberation of previously occupied territories, such as the now-infamous town of Bucha and the subsequent discovery of war crimes. Standing among the rubble in Bucha, Zelenskyy acknowledged "[it is] very important to us that the press is here . . . We want you to show the world what happened here. What the Russian military did. What the Russian Federation did in peaceful Ukraine. It was important for you to see that these were civilians."[101] In doing so, Ukraine is exercising two important levers against Russia that are identified by the literature on civilian resistance: (1) clearly delineating between violence perpetrated against forces versus noncombatants and (2) delegitimizing the oppressor by publicizing atrocities committed against civilians.[102] Both are critical in securing and maintaining external support and, by extension, inflicting additional costs on the occupier.

---

[98] Lisa Mascaro, "WATCH: 'We Need You Right Now,' Ukraine's Zelenskyy Asks Congress for More Help," *PBS*, March 16, 2022.

[99] Mascaro, 2022.

[100] Elena Moore, "Biden Pledges $800 Million to Ukraine After Zelenskyy's Plea for More U.S. Aid," *NPR*, March 16, 2022.

[101] Niamh Kennedy, "Zelensky Says Ukraine Wants to 'Show the World' What Happened in Bucha," *CNN*, blog, April 4, 2022.

[102] Binnendijk and Kepe, 2021, pp. 24–25.

Maintaining and Expanding Popular Support for Defense and Resistance

Lastly, scholarship on civilian-based resistance has identified the capacity of resistance movements to cultivate and maintain popular support for their cause as an important factor in their success. In fact, Binnendijk and Kepe have observed that "qualitative and quantitative analysis of nonviolent civil resistance movements suggests that active participation by a broad-based national coalition involving diverse groups and communities is correlated with the achievement of successful resistance outcomes."[103] By the same token, research on this issue "indicates that a coordinated society driven by common and coherent objectives is likely to put up more-effective resistance than a society that is more atomized."[104]

Ukrainian efforts to keep up morale have taken many forms since the beginning of the war, from official government communications to organic grassroots activities. For example, one formal means is a sophisticated, muti-channel messaging campaign aimed at both internal and external audiences and run by civilian officials. This includes Zelenskyy's evening addresses directed at the Ukrainian public, which the president has consistently published on Telegram, YouTube, and his office's website. Since the first days of the conflict, Zelenskyy traded in his business suits for an army green T-shirt and cargo pants, which he wears during his nightly addresses. Without uttering a word, Zelenskyy's choice to don "the single most accessible garment around—the T-shirt—serves as clear a statement of solidarity with his people as any of his rhetoric."[105]

The backdrops of Zelenskyy's addresses have been equally significant. When embassies and companies were moving personnel out of Kyiv and concerns about the president's safety mounted, Zelenskyy was careful to demonstrate that he was staying in the capital by posting videos from the

[103] Binnendijk and Kepe, 2021, p. 28.

[104] Binnendijk and Kepe, 2021, pp. 28–29.

[105] Vanessa Friedman, "The Man in the Olive Green Tee," *New York Times*, March 21, 2022.

city.[106] When Russian airstrikes rained down on Ukrainian infrastructure, including the power grid, Zelenskyy delivered his address from the pitch-black streets of Kyiv.[107] Similar to his wartime outfit, these backdrops convey a message of unity with the Ukrainian public. However, Zelenskyy is not alone in this approach: Officials across his administration have mirrored these methods in their communications to the Ukrainian public.

Ukrainian officials have also relied on the power of labels and lexicon to mobilize popular support. In their content propagated by social and traditional media, officials have been disciplined in branding Russian forces and decisionmakers as "occupiers." In cases in which Russian forces have clearly targeted Ukrainian civilians, officials have classified Russian forces as "terrorists," "war criminals" and "Rushists," a portmanteau combining "Russians" and "fascists."[108] These labels serve several purposes: (1) they deindividualize Russian soldiers, instead painting the military as a nameless, faceless mass; (2) they serve as an ongoing reminder of the fate that will befall Ukrainians should Russia prevail in its military campaign—the occupation of Ukraine and its people; and (3) the terms evoke painful, deep-seated historical memories of the Ukrainian experience in World War II (discussed in Chapter 2).

The efforts of Ukrainian authorities described above are only a snapshot of their robust attempts to preserve the Ukrainian people's resolve to resist. Individual officials and government agencies alike have participated in what appear to be coordinated communication campaigns. Their approach includes the use of synchronized messages across organizations and platforms, such as the daily tallies of losses suffered by the Russian Armed Forces that government ministries post via social media along with infographics.[109]

[106] Matthias Williams, "Smartphone in Hand, Ukraine's President Takes Centre Stage in a Capital Under Attack," Reuters, updated March 1, 2022.

[107] Sophia Ankel, "Zelenskyy Gave His Nightly Address from a Pitch-Dark Kyiv as Russian Strikes Force Blackouts on Ukraine," *Business Insider*, October 28, 2022.

[108] Office of the President of Ukraine, "Struggle Becomes Victory. And Ukraine Becomes United! Address of Volodymyr Zelenskyy 27.10.2022," video, October 27, 2022.

[109] See, for instance, Ministry of Foreign Affairs of Ukraine [@MFA_Ukraine], "256 Days of Full-Scale Russia's War on #Ukraine. Information on #Russian invasion. Losses of #Russia's armed forces in Ukraine, November 6," Twitter post, November 6, 2022;

Having adopted the framework developed by Binnendijk and Kepe, which classifies activities carried out by official civilian government entities as civilian-based resistance, we chose to include the messaging efforts described above in our analysis. That said, we acknowledge that some conceptions of resistance may dispute the inclusion of information campaigns carried out at the hands of the commander in chief and his coterie as civilian-based resistance. In either case, it is possible that the involvement of senior officials enables extra-governmental civilian-based resistance informational efforts aimed at preserving Ukrainian will to resist. However, further research is required to tease out the nuances of this relationship.

In addition to the efforts of civilian governmental officials, there are the prolific grassroots efforts of ordinary citizens and nongovernmental groups. Countless videos and images depicting Ukrainian bravery, loss, and humor have been captured and posted to social media over the course of the conflict. On the first day of the war, for instance, Ukrainians saw an older woman confront Russian soldiers in the street questioning, "What the F*** are you doing on our land with all these guns?" Before she could be shooed away, the video captured her attempting to hand the soldiers sunflower seeds, while explaining, "take these and put them in your pockets so at least sunflowers will grow when you all lie down here" to die.[110] The video went viral, garnering two million views on the first day. Other images, such as those of Ukrainian tractors towing abandoned Russian military equipment, have also spread like wildfire.[111]

The popularity of these messages could be taken as evidence of official and organic attempts at maintaining Ukrainian popular morale. However, the existence of activities aimed at cultivating support for the Ukrainian cause at home and eroding support for Russia's cause does not necessarily guarantee their success. This begs the question, what does existing evidence

---

and Ministry of Defense of Ukraine [@DefenceU], "'Good morning, my neighbors!' Prince Akeem Joffer Total combat losses of the enemy from Feb 24 to Nov 6: Image," Twitter post, November 6, 2022.

[110] Shweta Sharma, "Brave Ukrainian Woman Tells Russian Soldier: 'Put Sunflower Seeds in Your Pocket so They Grow When You Die,'" *Independent*, February 25, 2022.

[111] Chris Brown, "Famous for Towing Captured Russian Tanks, Ukrainian Farmers Step Up for War Effort," *CBC*, March 18, 2022.

tell us about the effectiveness of civilian-based efforts intended to bolster morale in Ukraine? Before attempting to answer this question, we must first consider the strength of popular support for Ukraine's defensive operations against Russia. Polling in Ukraine in the months since the initial Russian invasion on February 24, 2022, could offer some important clues. A survey conducted as early as late March–early April 2022 found that 97 percent of the respondents believed that Ukraine would win the war.[112] That figure remained constant in an analogous survey conducted in June 2022.[113] When a survey asked "do you believe in Ukraine's victory in the war against the Russian Federation" in Western Ukraine in May 2022, 94 percent of the respondents answered "yes" or "likely yes."[114] These figures suggest that, at least among those populations included in the samples, Ukrainian popular support for resisting Russian occupation has remained high. Even so, it is still unclear whether high Ukrainian morale can be attributed to Ukrainian resistance activities.

There are several factors that make it difficult to ascertain the role of Ukrainian civilian efforts in sustaining morale. First, the Kremlin's behavior in Ukraine has likely strengthened Ukrainian will to fight and resist. Russia's large-scale military assault, its indiscriminate shelling of civilian targets, and early reports of war crimes no doubt soiled Moscow's reputation in Ukraine, even among the country's most sympathetic audiences. In other words, Russia has tarnished its image without the Ukrainian resistance so much as lifting a finger. Relatedly, much of the world has rallied behind the Ukrainian cause, which could also be a factor in boosting Ukrainian will to resist. To summarize, the evidence available at the time of this writing cannot definitively attribute the high levels of Ukrainian civilian resolve to resist Russia's invasion since February 2022 to the efforts of the resistance.

---

[112] Center for Insights in Survey Research, "Public Opinion Survey of Residents of Ukraine, March 30–April 2, 2022" International Republican Institute, May 6, 2022, p. 8.

[113] Center for Insights in Survey Research, 2022, pp. 8–9.

[114] Ilko Kucheriv, Democratic Initiatives Foundation, "How the War Changed the Way Ukrainians Think About Friends, Enemies, and the Country's Strategic Goals," May 30, 2022.

CHAPTER 4

Conclusions

Civilian-based resistance throughout the first four months of the 2022 Russian invasion of Ukraine included a variety of activities and actors. Individual activists, self-organized groups, social actors, enterprises, and civilian government institutions engaged in activities aimed at helping Ukrainian civilians in war-affected areas, confronting Russian soldiers on the ground or in cyberspace, offering various types of support to the Ukrainian military efforts, and engaging in information-based activities. Often, civilian-based activities have been spontaneous and need-based, and they have relied on the existing informal networks at the local community level. According to one Ukrainian participant of a self-organized group, "during the war, the community has strengthened its capacity for self-organization, it acts as a defensive barrier . . . as if it were a colony of ants."[1] Russian forces failed to overwhelm the Ukrainian territory during the first four months of war, and the Ukrainian government continued to function. Despite the government's continued ability to function, according to Felip Daza Sierra, "the vast majority of nonviolent actions have been organized locally without coordination at the national level."[2] State-level or supra-local-level coordination has been present in the areas of "civil protection, humanitarian aid, war crimes monitoring, mass nonviolent communications, and hacker-activism."[3]

The Ukrainian example may not serve as a perfect case for the study of civilian-based resistance against an external occupier in the traditional sense, in which an overwhelming part of the territory is occupied by the

[1] Daza Sierra, 2022.

[2] Daza Sierra, 2022.

[3] Daza Sierra, 2022.

external power, and the country's government is forced to abandon its seat. After all, at the peak of their land grab, Russian forces had invaded only 25 percent of Ukrainian territory, and the Ukrainian government has continued to govern without interruption.[4] Even so, our initial exploration of the role of civilians in the ongoing conflict indicates that this area is ripe with lessons for future analogous conflicts.

The war in Ukraine highlights several trends that might be relevant to future civilian-based resistance campaigns against foreign-occupying powers (summarized in Table 4.1):

1. The war in Ukraine provides a glimpse into what future cyber wars could look like. Civil resistance scholar Michael Beer posits that the proliferation of digital technologies in the 21st century could offer civilians more opportunities to engage in cyber-based activities in support of a resistance movement.[5] Inter-state conflict could involve large-scale cyber operations waged by a loose confederation of amateurs and professionals from the occupied country and abroad. Countries endowed with robust cyber capabilities—whether state-affiliated or independent—but weaker conventional forces could leverage their technical skills to create an asymmetric advantage.[6] However, the initial months of the Russia-Ukraine War (2022) also illustrate that cyber operations would require "sustained and systematic" efforts to be effective against an authoritarian regime such as Russia, in which society has little ability to affect the governmental decisions.[7]
2. Civilians with specialist or generalist skills may offer meaningful contributions by helping to impose direct military costs on occupying powers or by supporting the defending armed forces or civil

[4] Pierre Breteau, "Nine Months of War in Ukraine in One Map: How Much Territory Did Russia Invade and Then Cede?" *Le Monde*, November 25, 2022.

[5] Michael A. Beer, *Civil Resistance Tactics in the 21st Century*, International Center on Nonviolent Conflict, 2021.

[6] Lorenzo Franceschi-Bicchierai, "Inside Ukraine's Decentralized Cyber Army," *Vice*, July 19, 2022.

[7] Lewis, 2022.

society. Literature on national resistance against foreign occupation often highlights the need to teach civilians tactical skills that are directly relevant to the military fight.[8] However, the Ukraine case illustrates how nonviolent civilian-based resistance "sets a low threshold for individual and collective participation and, thus, offers opportunities for many groups to join."[9] Regardless of defense or security-related expertise, people of any age can still contribute by assisting their fellow civilians (particularly those who are more vulnerable), creating networks to distribute food and medicine, assisting internally displaced people and refugees, or helping to boost their counterparts' morale. It is therefore necessary to prepare guidelines for such eventualities and avoid instances in which a sudden surge of civilian volunteers creates chaos, causes friendly fire, or interferes with the strategic objectives of the country.

3. The ability to deny an occupier's economic consolidation will increasingly signify protecting not only economic centers of gravity during the war but also peacetime resilience preparations. This includes diversifying energy resources, other product supplies, and—if the country is a significant exporter—export lines, especially in unoccupied territories. These efforts could help prevent the occupier's consolidation of economic control and support civilians by continuing economic activity in unoccupied territories.
4. Although the shock of an external military invasion may provoke the mobilization of mass civil-resistance movements, the Ukraine case indicates that engaging the public in civil society before a conflict is also important for building wartime resistance movements. Prior experiences may contribute to what Maciej Bartkowski calls "the diffusion of the civil resistance know-how," whereby "with each victory—and failure—popular resisters learn from experiences of

[8] Marta Kepe and Jan Osburg, "Total Defense: How the Baltic States Are Integrating Citizenry into Their National Security Strategies," *Small Wars Journal*, September 24, 2017.

[9] Maciej J. Bartkowski, "Insights into Nonviolent Liberation Struggles," in Maciej J. Bartkowski, ed., *Recovering Nonviolent History: Civil Resistance in Liberation Struggles*, Lynne Rienner, 2013, p. 347.

their own as well as those of others."[10] The in-depth study of the history of national civil resistance movements of countries located in potential conflict areas might help the United States and its allies and partners better understand how civil resistance is conceptualized in these countries, trends in civil resistance methods, and their application to current or future conflicts.

5. Multinational companies and large businesses have played various roles during wars. In Ukraine, large companies have become involved either intentionally (e.g., donating humanitarian aid or supporting the resistance or military defense financially or in-kind), unintentionally (via online sales or property rental sites used to promote services in support of the Ukrainian government, its armed forces, citizens, or business entities), or as a result of public pressure (e.g., public shaming campaigns aimed at companies that operate in Russia).[11] In future wars, large companies could increasingly shape the operational aspects of a conflict (for instance, by providing new or enhanced capabilities to improve reconnaissance) and the geopolitical and strategic outcome of a conflict.[12]
6. The civilian-based resistance activities in Ukraine offer a valuable learning experience for the United States and its allies in terms of understanding the adversary. Insights from the war in Ukraine may help the United States and its allies continue to readjust away from the decades-long focus on counterinsurgency operations towards great power conflicts.[13]

It should be noted that civilian-based resistance in Ukraine has benefited from several preexisting factors: (1) Ukraine is only partially occupied, which offers territory to host Ukraine's government, resistance-

[10] Bartkowski, 2013, p. 346.

[11] Cat Zakrzewski, "Pressure Mounts on Major Tech Companies to Take Tougher Line Against Russia," *Washington Post*, February 26, 2022.

[12] Abushur Prakash, "How Technology Companies Are Shaping the Ukraine Conflict," *Scientific American*, October 28, 2022.

[13] Think JSOU, "ThinkJSOU Panel: Update on the Ukraine / Russia Conflict," video, October 14, 2022.

TABLE 4.1
Summary of Ukrainian Resistance Efforts by Proximate Objective

Proximate Objective	Characterization
Impose direct or indirect costs on occupying force	Activities include a variety of civilian-based actions that may impose direct or indirect costs on the Russian forces, either through support to the armed fight (joining the Ukrainian armed formations) and military effort via the contribution of specific skills (cyber, OSINT) or through engagement in nonviolent protests and seeking to physically impede Russian troop movements.
Secure external support	A priority area for Ukrainian civilian-based activities has been to secure the continuous support of external actors: foreign governments, international organizations, enterprises, and individuals. Ukraine also used more-traditional civilian-based resistance activities in combination with new technology and activities in the information environment, and Ukraine has asked external actors to help impose costs on Russia via economic and political means.
Deny occupier's political and economic consolidation	There has been a strong emphasis on ensuring the continuity of government and messaging around a stable Ukrainian head of state, severing diplomatic relations with Russia, and undermining Russian-organized regional referenda. The war underlines the importance of building food and energy resilience prior to a conflict.
Reduce occupier's capacity for repression	Activities include volunteer grassroots efforts to protect and assist civilians and displaced persons inside and outside Ukraine and to raise awareness of the atrocities committed against noncombatants.
Maintain and expand popular support for defense and resistance	Activities include a sophisticated, multi-channel messaging campaign aimed at both internal and external audiences; using the power of labels and lexicon to mobilize popular support; and prolific grassroots efforts publicizing Ukrainian solidarity and civilian efforts.

coordinating structures, resistance supporters, and sanctuary for internally displaced people; (2) Ukraine's unoccupied territories share long borders with western countries, which allow for multiple avenues for refugee flows and the inflow of external support; (3) Ukraine has taken advantage of modern technologies (such as IT, satellite communications, and cell phones) to support the military defense effort, political and economic activities, and civilians; and (4) Ukraine has built on the momentum of an increasingly active civil society that was inspired by the 2013–2014 Euromaidan protests and the 2014 Russia–Ukraine War in Eastern Ukraine.[14] The shock of war and Putin's fervent speech announcing the "special operation" in Ukraine contributed to massive volunteerism and societal cohesion in the face of a common threat.

At the same time, resistance may have been hampered and may be negatively affected in the future by the following factors: (1) peacetime corruption, governance issues, and gaps in critical infrastructure resilience and civil defense systems might affect the efficiency and morale of civilian-based resistance as the war drags on; (2) the national political leadership's messaging in the weeks and days immediately preceding the war might have undermined civil society's ability to better prepare for war, and other preparations might have been too little too late; and (3) some wartime resistance activities might have suffered from poor coordination among the multitude of actors, and others might hang in the balance in terms of their adherence to international law or norms. This might cause a backlash in later stages of the war or during the peacebuilding and reconstruction stage.

This report is intended to be a high-level overview of some aspects of Ukrainian civilian-based resistance activities during the first four months of the war. Therefore, a more-detailed study of the topic will be required, particularly as more information becomes available on other aspects of the war and the impact of civilian activities on strategic and military decisions.

---

[14] Ritu Nayyar-Stone, "Ukraine Civil Society Assessment," NORC at the University of Chicago, undated.

Abbreviations

AI	artificial intelligence
DDoS	distributed denial-of-service
EU	European Union
IT	information technology
MOD	Ministry of Defense of Ukraine
OSCE	Organization for Security and Co-Operation in Europe
OSINT	open-source intelligence
OUN	Organization of Ukrainian Nationalists
SBU	Security Service of Ukraine
UN	United Nations

References

ABC News, "Protesting Civilians Confront Russian Military Vehicles in Ukraine," video, March 1, 2022. As of August 10, 2022: https://www.youtube.com/watch?v=zcJZ-9plJ-o

Abdelaziz, Salma, "As Russian Attacks Continue to Pound Ukraine, Volunteers Are Rebuilding Kyiv's Suburbs," *CNN*, June 13, 2022.

Adams, Paul, "'Shame on You': How President Zelensky Uses Speeches to Get What He Needs," *BBC*, March 24, 2022.

Aivazovska, Olga, "Pseudo-Referendums in the Invaded Territories of Ukraine: What Russia Plans to Do and How to Disable Them," Opora, August 10, 2022.

Ankel, Sophia, "Zelenskyy Gave His Nightly Address from a Pitch-Dark Kyiv as Russian Strikes Force Blackouts on Ukraine," *Business Insider*, October 28, 2022.

Applebaum, Anne, *Red Famine: Stalin's War on Ukraine*, Doubleday, 2017.

Archick, Kristin, "Russia's Invasion of Ukraine: European Union Responses and Implications for U.S.-EU Relations," Congressional Research Service, updated July 28, 2022.

Arraf, Jane, "To House Refugees, Lviv Wants to Make Beautiful Buildings That Last," *New York Times*, May 31, 2022, updated June 2, 2022.

Bartkowski, Maciej J., "Insights into Nonviolent Liberation Struggles," in Maciej J. Bartkowski, ed., *Recovering Nonviolent History: Civil Resistance in Liberation Struggles*, Lynne Rienner, 2013.

Bartkowski, Maciej J., "Countering Hybrid War: Civil Resistance as a National Defence Strategy," *OpenDemocracy*, May 12, 2015.

Becatoros, Elena, "Ukraine: Drivers Risk All to Bring Aid, Help Civilians Flee," Associated Press, June 10, 2022.

Beer, Michael A., *Civil Resistance Tactics in the 21st Century*, International Center on Nonviolent Conflict, 2021.

Bellini, Jason, "Ukrainian Civilians Use Drones to Help Fight Against Russian Forces," *Scripps News*, May 3, 2022.

Binnendijk, Anika, "Holding Fire: Security Force Allegiance During Nonviolent Uprisings," thesis, Fletcher School of Law and Diplomacy, Tufts University, August 2009.

Binnendijk, Anika, and Marta Kepe, *Civilian-Based Resistance in the Baltic States: Historic Precedents and Current Capabilities*, RAND Corporation, RR-A198-3, 2021. As of February 28, 2022:
https://www.rand.org/pubs/research_reports/RRA198-3.html

Blinken, Antony J., "Secretary Blinken with Margaret Brennan of CBS News," U.S. Department of State, March 6, 2022a.

Blinken, Antony J., "Additional Humanitarian Assistance for the People of Ukraine," U.S. Department of State, July 9, 2022b.

Bloomberg Quicktake, "Life in a Shipping Container: Ukraine's Refugees in Lviv," video, September 19, 2022. As of August 10, 2022:
https://www.youtube.com/watch?v=qVV8_OTt1ms

Breteau, Pierre, "Nine Months of War in Ukraine in One Map: How Much Territory Did Russia Invade and then Cede?" *Le Monde*, November 25, 2022.

Brown, Chris, "Famous for Towing Captured Russian Tanks, Ukrainian Farmers Step Up for War Effort," *CBC*, March 18, 2022.

Burgess, Matt, "Ukraine's Volunteer 'IT Army' Is Hacking in Uncharted Territory," *Wired*, February 27, 2022.

Butchenko, Maksym, "Ukraine's Territorial Defence on a War Footing," International Centre for Defence and Security, April 13, 2022.

Bykowicz, Julie, and Vivian Salama, "Ukraine Lobbyists in U.S. Pivot from Pipeline to War Aid and Sanctions," *Wall Street Journal*, March 23, 2022.

Cabinet of Ministers of Ukraine, "Розпорядження 'Про Затвердження Плану Заходів Забезпечення Продовольчої Безпеки В Умовах Воєнного Стану' [Order 'On the Approval of the Plan of Measures to Ensure Food Security in the Conditions of Martial Law']," No. 327, April 29, 2022.

Center for Insights in Survey Research, "Public Opinion Survey of Residents of Ukraine, March 30–April 2, 2022," International Republican Institute, May 6, 2022. As of August 18, 2022:
https://www.iri.org/resources/public-opinion-survey-of-residents-of-ukraine/

Chernichkin, Kostyantyn, and Anton Protsiuk, "Vostok SOS Helps Evacuate over 23,000 People Fleeing Russia's War, Continues Its Mission," *Kyiv Independent*, July 25, 2022.

Collins, Liam, and John Spencer, "In Ukraine, Two US Army Veterans See the New Face of War," *Soldier of Fortune Magazine*, August 22, 2022.

"Coordination Headquarters for Treatment of Prisoners of War Established in Ukraine," *Ukrinform*, March 12, 2022.

Costello, Norma, and Vera Mironova, "Ukraine Has a Secret Resistance Operating Behind Russian Lines," *Foreign Policy*, November 21, 2022.

Courtney, William, and Khrystyna Holynska, "Continuity of Government in Ukraine," *RAND Blog*, February 25, 2022. As of May 16, 2022:
https://www.rand.org/blog/2022/02/continuity-of-government-in-ukraine.html

Crimea SOS, "КримSOS: В Донецькій області розповіли про шляхи повернення українців з рф [CrimeaSOS: In the Donetsk Region, They Talked About Ways to Return Ukrainians from the Russian Federation]," April 22, 2022.

Dabrowski, Marek, "Ukraine's Oligarchs Are Bad for Democracy and Economic Reform," *Bruegel*, blog, October 3, 2017.

D'Anieri, Paul, *Ukraine and Russia: From Civilized Divorce to Uncivil War*, Cambridge University Press, 2019.

Daza Sierra, Felip, "Ukrainian Nonviolent Civil Resistance in the Face of War: Analysis of Trends, Impacts and Challenges of Nonviolent Action in Ukraine Between February and June 2022," International Catalan Institute for Peace and International Institute for Nonviolent Action, 2022.

Defense of Ukraine [@DefenceU], "We will rebuild everything that was destroyed. Our spirit will remain unbroken," Twitter post, November 3, 2022. As of May 16, 2022:
https://twitter.com/DefenceU/status/1588170601058865152

Diuk, Nadia, "EUROMAIDAN: Ukraine's Self-Organizing Revolution," *World Affairs*, Vol. 176, No. 9, March–April 2014.

Ekman, Ivar, and Per-Erik Nilsson, "Folkets underrättelsetjänst—öppna källor 'OSINT' och Ukraina [The People's Intelligence Service—Open Source 'OSINT' and Ukraine]," Swedish Defence Research Agency, project number 1498503, June 2022.

Embassy of Ukraine in the Russian Federation, homepage, undated. As of September 2, 2022:
https://russia.mfa.gov.ua

Evropeiska Pravda, "МЗС Повідомляє Про Припинення Роботи Усіх Консульств України В Росії [The Ministry of Foreign Affairs Announces the Suspension of All Ukrainian Consulates in Russia]," March 13, 2022.

Fenton, Nick, and Andrew Lohsen, "Corruption and Private Sector Investment in Ukraine's Reconstruction," Center for Strategic and International Studies, November 8, 2022.

Fiala, Otto, *Resistance Operating Concept*, Swedish Defense University and Special Operations Command Europe, 2019.

Franceschi-Bicchierai, Lorenzo, "Inside Ukraine's Decentralized Cyber Army," *Vice*, July 19, 2022.

Friedman, Vanessa, "The Man in the Olive Green Tee," *New York Times*, March 21, 2022.

Goldstein, Joshua, "The Role of Digital Networked Technologies in the Ukrainian Orange Revolution," Berkman Center for Internet and Society, Harvard University, Research Publication No. 2007-14, December 2007.

Government of Ukraine, "Конституція України—Розділ V [Constitution of Ukraine—Chapter V]," undated-a. As of September 1, 2022:
https://www.president.gov.ua/ua/documents/constitution/konstituciya-ukrayini-rozdil-v

Government of Ukraine, "Donate to Ukraine's Defenders," webpage, undated-b. As of August 30, 2022:
https://war.ukraine.ua/donate

Grey, Stephen, Tom Bergin, Sevgil Musaieva, and Roman Anin, "Putin's Allies Channelled Billions to Ukraine Oligarch," Reuters, November 26, 2014.

Hautala, Laura, "Big Tech's Support for Ukraine Recasts Industry's Global Role," *CNET*, March 1, 2022.

Hromadske Int. [@Hromadske], "Grand Theft Russian Military Vehicles: Ukraine," Twitter post, March 6, 2022. As of May 17, 2022:
https://twitter.com/hromadske/status/1500603614980722689

Ilko Kucheriv Democratic Initiatives Foundation, "How the War Changed the Way Ukrainians Think About Friends, Enemies, and the Country's Strategic Goals," May 30, 2022.

IT Army of Ukraine, "IT ARMY of Ukraine," Telegram post, May 18, 2022. As of August 18, 2022:
https://t.me/itarmyofukraine2022/364 or https://archive.ph/KMKTj

Izbirkom, "Хроніка Вторгнення Росія Затвердила Введення Військ На Схід України [The Chronicle of the Invasion of Russia Approved the Introduction of Troops to the East of Ukraine]," April 1, 2022. As of August 19, 2022:
https://izbirkom.org.ua/publications/politika-8/2022/hronika-vtorgnennya-rosiya-zatverdila-vvedennya-vijsk-na-shid-ukrayini

Kamenetz, Anya, "Shelter.Lviv Started on Instagram. It's Now Helped House 4,000 Women and Children," *NPR*, May 17, 2022.

Karatnycky, Adrian, "The Fall and Rise of Ukraine's Political Opposition: From Kuchmagate to the Orange Revolution," in Ander Åslund and Michael McFaul, eds., *Revolution in Orange: The Origins of Ukraine's Democratic Breakthrough*, Carnegie Endowment for International Peace, 2006.

Kennedy, Niamh, "Zelensky Says Ukraine Wants to 'Show the World' What Happened in Bucha," *CNN*, blog, April 4, 2022.

Kepe, Marta, and Anika Binnendijk, "What Is Continuity of Government, and Why Does It Matter for Ukraine's National Resistance?" *RAND Blog*, March 17, 2022. As of June 1, 2022: https://www.rand.org/blog/2022/03/what-is-continuity-of-government-and-why-does-it-matter.html

Kepe, Marta, and Jan Osburg, "Total Defense: How the Baltic States Are Integrating Citizenry into Their National Security Strategies," *Small Wars Journal*, September 24, 2017.

"Kherson: How Is Russia Imposing Its Rule in Occupied Ukraine?" *BBC*, May 11, 2022.

Kurtzleben, Danielle, "Volunteer Hackers Form 'IT Army' to Help Ukraine Fight Russia," transcript of interview with Dina Temple-Raston, *NPR*, March 27, 2022.

Kuzio, Taras, "Everyday Ukrainians and the Orange Revolution," in Ander Åslund and Michael McFaul, eds., *Revolution in Orange: The Origins of Ukraine's Democratic Breakthrough*, Carnegie Endowment for International Peace, 2006.

Kvartal 95, "About Us," webpage, undated. As of August 11, 2022: https://kvartal95.com/en/about

Labott, Elise, "'We Are the First in the World to Introduce This New Warfare': Ukraine's Digital Battle Against Russia," *Politico*, March 8, 2022.

Lambert, Jesse [@watchmans_way], "Russia sneaks into the country using civilian vehicles. This one clearly marked Z is bum rushed by dozens of citizens," Twitter post, March 1, 2022. As of March 3, 2022: https://twitter.com/watchmans_way/status/1498602914541154306

Lapshin, Oleksandr, "Як мешканці одного будинку стали волонтерити разом. Історії з Харкова [How Residents of the Same House Began to Volunteer Together: Stories from Kharkiv]," *Ukrainska Pravda*, April 17, 2022.

Lazar, Nomi Claire, "Need an Expert? War in Ukraine: President Volodymyr Zelensky's Rhetoric," University of Ottawa, March 22, 2022.

LB.ua, "SSU, Return Alive Foundation and Volunteers Launch Joint Service," March 2, 2022.

Legistorm, "Senate Ukraine Caucus," webpage, undated. As of July 12, 2022: https://www.legistorm.com/organization/summary/128483/Senate_Ukraine_Caucus.html

Lewis, James A., "Cyber War and Ukraine," Center for Strategic and International Studies, June 2022.

Liber, George O., *Total Wars and the Making of Modern Ukraine, 1914–1954*, University of Toronto Press, 2016.

Linnane, Ciara, "Companies That Exited Russia After Its Invasion of Ukraine Are Being Rewarded with Outsize Stock-Market Returns, Yale Study Finds—and Those That Stayed Are Not," *Market Watch*, June 8, 2022.

Mascaro, Lisa, "WATCH: 'We Need You Right Now,' Ukraine's Zelensky Asks Congress for More Help," *PBS*, March 16, 2022.

Matthews, Chris, "Sending Putin's Invaders into Reverse: Defiant Ukrainians Chant 'Go Home' as They Force Back Two Russian Military Vehicles Marked with a 'Z' During Fearless Demonstration in Kherson," *Daily Mail*, March 20, 2022.

McConnell, Patrick J., "In Ukraine's Lviv, Help—and Maybe Hope—for the Legions Displaced by War," *Los Angeles Times*, March 24, 2022.

McFaul, Michael, "Ukraine Imports Democracy: External Influences on the Orange Revolution," *International Security*, Vol. 32, No. 2, Fall 2007.

MilitaryLand.net [@Militarylandnet], "Civilians blocking the passage of another Russian column in #Melitopol #Ukraine #UkraineRussiaWar," Twitter post, March 1, 2022. As of May 11, 2022:
https://twitter.com/Militarylandnet/status/1498645822879186954

Ministry of Agrarian Policy and Food of Ukraine, "Тарас Висоцький: Україна сформувала продовольчі резерви для своїх громадян [Taras Vysotsky: Ukraine Has Formed Food Reserves for Its Citizens]," July 22, 2022. As of August 18, 2022:
https://www.kmu.gov.ua/news/taras-vysotskyi-ukraina-sformuvala-prodovolchi-rezervy-dlia-svoikh-hromadian

Ministry of Defense of Ukraine [@DefenceU], "'Good morning, my neighbors!' Prince Akeem Joffer Total combat losses of the enemy from Feb 24 to Nov 6: Image," Twitter post, November 6, 2022. As of November 6, 2022:
https://twitter.com/DefenceU/status/1589204192975097856

Ministry of Economy of Ukraine, "Аудит "Держрезерву" Підтвердив Тотальне Розкрадання Запасів І Мільйонні Збитки, ["The Audit of 'State Reserve' Confirmed the Total Theft of Stocks and Millions of Losses]," November 5, 2020. As of April 18, 2022:
https://www.me.gov.ua/News/Detail?lang=uk-UA&id=d0838d5d-6d86-4aa0-82e0-58f369836994&title=AuditderzhrezervuPidtverdivTotalneRozkradanniaZapasivIMilionniZbitki

Ministry of Foreign Affairs of Ukraine, "Ukrainians Worldwide," webpage, December 18, 2019. As of April 11, 2022:
https://mfa.gov.ua/en/about-ukraine/ukrainians-worldwide

Ministry of Foreign Affairs of Ukraine, "Організація з безпеки і співробітництва в Європі (ОБСЄ) [Organization for Security and Cooperation in Europe (OSCE)]," July 20, 2022.

Ministry of Foreign Affairs of Ukraine [@MFA_Ukraine], "256 Days of Full-Scale Russia's War on #Ukraine. Information on #Russian invasion. Losses of #Russia's armed forces in Ukraine, November 6," Twitter post, November 6, 2022. As of November 6, 2022: https://twitter.com/MFA_Ukraine/status/1589201979808837632

Ministry of Youth and Sports of Ukraine, "Українські Спортсмени Жертвують Своїми Найціннішими Нагородами Для Захисту України Від Російської Агресії Та Змінюють Спортивну Форму На Військову [Ukrainian Athletes Donate Their Most Valuable Awards to Protect Ukraine from Russian Aggression and Change Their Sports Uniforms to Military Uniforms]," April 6, 2022.

Moore, Elena, "Biden Pledges $800 Million to Ukraine After Zelenskyy's Plea for More U.S. Aid," *NPR*, March 16, 2022.

Mukherjee, Pradipta, "Russian Soldiers Offered Bitcoin in Exchange for White Flag," *Yahoo Finance*, February 27, 2022.

Nayyar-Stone, Ritu, "Ukraine Civil Society Assessment," NORC at the University of Chicago, undated.

Nepstad, Sharon Erickson, *Nonviolent Revolutions: Civil Resistance in the Late 20th Century*, Oxford University Press, 2011.

New Format, "У Києві Затвердили План Евакуації Населення На Випадок Надзвичайної Ситуації [In Kyiv, an Emergency Evacuation Plan Was Approved]," February 12, 2022. As of April 6, 2022: https://newformat.info/news/u-kiievi-zatverdili-plan-evakuacii-naselennya-na-vipadok-nadzvichajnoi-situacii

Nichols, Michelle, "U.N. Suspends Russia from Human Rights Body, Moscow Then Quits," Reuters, April 7, 2022.

O'Connor, Coilin, and Halyna Tereshchuk, "The Revolution on Granite: Ukraine's 'First Maidan,'" *Radio Free Europe/Radio Liberty*, October 15, 2020.

Office of the President of Ukraine, "Struggle Becomes Victory. And Ukraine Becomes United! Address Of Volodymyr Zelenskyy 27.10.2022," video, October 27, 2022. As of April 14, 2023: https://www.youtube.com/watch?v=LQ9yobovFmw

Omelchenko, Volodymyr, "Базові причини енергетичної кризи в ОЗП 2021–2022pp [Basic Causes of the Energy Crisis in the 2021–2022 OZP]," Razumkov Centre, November 24, 2021. As of April 6, 2022: https://razumkov.org.ua/statti/bazovi-prychyny-energetychnoi-kryzy-v-ozp-2021-2022rr

Opora, "Як львівський суд забороняв проросійські політичні партії. Аналіз 16 судових процесів [How the Lviv Court Banned Pro-Russian Political Parties. Analysis of 16 Trials]," August 17, 2022. As of August 18, 2022:
https://www.oporaua.org/news/vybory/partii/24260-iak-lvivskii-sud-zaboroniav-prorosiiski-politichni-partiyi-analiz-16-sudovikh-protsesiv

Organization for Security and Co-Operation in Europe, "Moscow Mechanism," December 1, 1991.

Panasiuk, Kateryna, and Mykyta Vorobiov, "Charity Begins with Drones: Funding Ukraine's Resistance," Center for European Policy Analysis, September 28, 2022.

Polish Investment and Trade Agency, "The Polish Investment and Trade Agency (PAIH) Provides Free Office Space for Ukrainian Companies," March 1, 2022.

Prakash, Abishur, "How Technology Companies Are Shaping the Ukraine Conflict," *Scientific American*, October 28, 2022.

Prasad, Ajnesh, "Volodymyr Zelensky's Appeal Lies in His Service to Ukrainians Above All Else," *The Conversation*, March 2, 2022.

Price, Ned, "Invocation of the OSCE Moscow Mechanism to Investigate Mounting Reports of Human Rights Abuses and International Humanitarian Law Violations by Russia in Ukraine," press statement, U.S. Department of State, June 3, 2022.

Prokip, Andrian, "Stability Against All Odds: Ukrainian Energy During the First Weeks of the War," *Wilson Center*, blog, March 28, 2022.

Prus, Maria, "Як підтримка малих бізнесів може допомогти швидшому відродженню економіки України [How Supporting Small Businesses Can Help the Faster Revival of Ukraine's Economy]," *Ukrainian VOA*, May 4, 2022.

Prytula, Olena, "The Ukrainian Media Rebellion," in Ander Åslund and Michael McFaul, eds., *Revolution in Orange: The Origins of Ukraine's Democratic Breakthrough*, Carnegie Endowment for International Peace, 2006.

Puglisi, Rosaria, "A People's Army: Civil Society as a Security Actor in Post-Maidan Ukraine," Istituto Affari Internazionali, July 2015.

"Read Putin's National Address on a Partial Military Mobilization," *Washington Post*, September 21, 2022.

Roshchina, Olena, "Kyiv Traffic Is Moving Again at Strike Point Near Shevchenko Park and Repair Work Begins on 'Klychko Bridge,'" *Ukrainska Pravda*, October 11, 2022.

Rozhanskiy, Timofei, "Why Russian Soldiers Are Refusing to Fight in the War on Ukraine," *Radio Free Europe/Radio Liberty*, July 20, 2022.

Sachalko, Boris, and Andrey Shurin, "Офисы под землей и указатели в никуда. В каком состоянии находятся бомбоубежища в Украине [Offices Underground and Signs to Nowhere. In What Condition Are the Bomb Shelters in Ukraine]," *Current Time*, December 12, 2021.

Salerno-Garthwaite, Andrew, "OSINT in Ukraine: Civilians in the Kill Chain and Information Space," *Global Defence Technology*, October 2022.

Sauer, Pjotr, "'They Were Furious': The Russian Soldiers Refusing to Fight in Ukraine," *The Guardian*, May 12, 2022.

Schwirtz, Michael, Anton Troianovski, Yousur Al-Hlou, Masha Froliak, Adam Entous, and Thomas Gibbons-Neff, "Putin's War: The Inside Story of a Catastrophe," *New York Times*, December 16, 2022.

Sharma, Shweta, "Brave Ukrainian Woman Tells Russian Soldier: 'Put Sunflower Seeds in Your Pocket so They Grow When You Die,'" *Independent*, February 25, 2022.

Shveda, Yuriy, and Joung Ho Park, "Ukraine's Revolution of Dignity: The Dynamics of Euromaidan," *Journal of Eurasian Studies*, Vol. 7, No. 1, January 2016.

Sisk, Richard, "Captured Russian Troops Call Home While Filmed by Ukrainian Officials, Raising Geneva Convention Questions," *Military.com*, March 1, 2022.

Snyder, Timothy, *The Reconstruction of Nations: Poland, Ukraine, Lithuania, Belarus, 1569–1999*, Yale University Press, 2003.

Snyder, Timothy, *Bloodlands: Europe Between Hitler and Stalin*, Basic Books, 2010.

Soesanto, Stefan, *The IT Army of Ukraine: Structure, Tasking, and Ecosystem*, Center for Security Studies, ETH Zürich, June 2022.

Sonne, Paul, Isabelle Khurshudyan, Serhiy Morgunov, and Kostiantyn Khudov, "Battle for Kyiv: Ukrainian Valor, Russian Blunders Combined to Save the Capital," *Washington Post*, August 24, 2022.

"У Держрезерві заявляють про критично низькі запаси зерна [The State Reserve Declares Critically Low Grain Reserves]," *Ukrinform*, September 20, 2021.

Think JSOU, "ThinkJSOU Panel: Update on the Ukraine / Russia Conflict," video, October 14, 2022. As of June 1, 2023: https://www.youtube.com/watch?v=Sz3lC1vHNEI

Thomas, Natalie, "Volunteers Evacuate the Elderly from Ukraine's Bakhmut, Fearing Russian Advance," Reuters, June 28, 2022.

Tomson, Bill, "Negotiations Intensify to Save Deal on Ukraine Grain Exports," *Agri-Pulse*, November 14, 2022.

Tymoshenko, Kyrylo, "Бульвар Тараса Шевченка [Taras Shevchenko Boulevard]," Telegram post, October 11, 2022. As of April 14, 2023: https://t.me/tymoshenko_kyrylo/2269

"Ukraine Repairs Roads Smashed by Russian Missiles in Kyiv, Dnipro Overnight—PHOTOS," *Euromaidan Press*, October 11, 2022.

Ukrainian Congress Committee of America, "Senate Ukraine Caucus," webpage, undated. As of July 12, 2022:
https://ucca.org/caucuses/
?doing_wp_cron=1657651075.0309031009674072265625

Ukrainian Congress Committee of America, Illinois Division, "Letter to the Honorable Antony Blinken," Facebook post, August 2, 2022. As of August 3, 2022:
https://www.facebook.com/UCCAILLINOIS

Ukrainian Congress Committee of America, Illinois Division, Facebook page, undated. As of August 3, 2022:
https://www.facebook.com/UCCAILLINOIS/

"Ukrainian Resistance: Man Tries to Push Back Russian Tank—Video," *The Guardian*, February 26, 2022.

United Nations, *Geneva Convention Relative to the Treatment of Prisoners of War*, adopted August 12, 1949, by the Diplomatic Conference for the Establishment of International Conventions for the Protection of Victims of War, held in Geneva from April 21 to August 12, 1949.

U.S. Department of Energy, "Ukraine Launches Electricity Exports to European Union with Support from the U.S. Department of Energy," July 2, 2022.

Vaccaro, Dan, "'Islands of Hope': Ukrainian Libraries Respond to Russian Invasion in Surprising and Heartbreaking Ways—Here's How," Berkeley Library, University of California, April 7, 2022.

Voices of Children, "The Results of the Fund Work During of the Full-Scale War in Ukraine," September 4, 2022.

Vostok SOS, "About Us," webpage, undated. As of November 1, 2022:
https://vostok-sos.org/en/about/mission/

Weiss, Michael, "Inside Ukraine's Psyops on Russian and Belarusian Soldiers," *New Lines Magazine*, March 29, 2022.

"Why Ukraine's Richest Oligarch Gave Up His Media Empire," *Yahoo News*, July 13, 2022:
https://news.yahoo.com/why-ukraine-richest-oligarch-gave-140100114.html

Williams, Matthias, "Smartphone in Hand, Ukraine's President Takes Centre Stage in a Capital Under Attack," Reuters, updated March 1, 2022.

Wright, Robin, "Will Mercenaries and Foreign Fighters Change the Course of Ukraine's War?" *New Yorker*, April 5, 2022.

Zakrzewski, Cat, "Pressure Mounts on Major Tech Companies to Take Tougher Line Against Russia," *Washington Post*, February 26, 2022.

"Zelensky Tells UN That Russia Must Be Expelled from Security Council," *France 24*, April 5, 2022.

Zelenskyy, Volodymyr, "Recognition of Russia as Terrorist State Needed Not as Political Gesture, but as Effective Defense of Free World—Address of President of Ukraine," Office of the President of Ukraine, July 30, 2022.

Zhernakov, Mykhailo, "Ukraine's Vibrant Civil Society Deserves Key Role in Post-War Transformation," Atlantic Council, July 19, 2022.